CW00543419

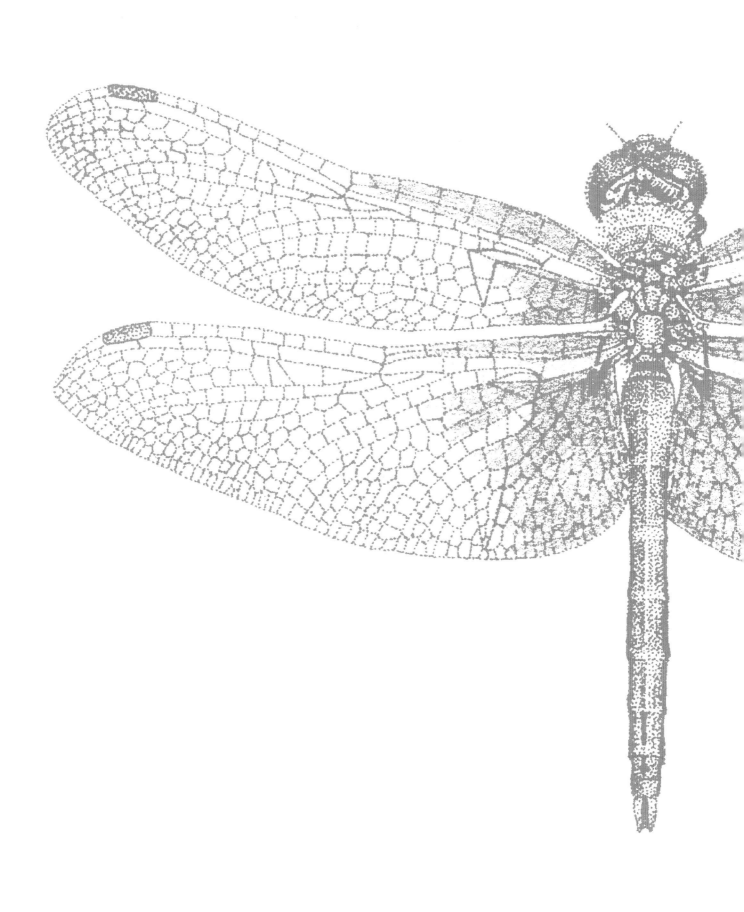

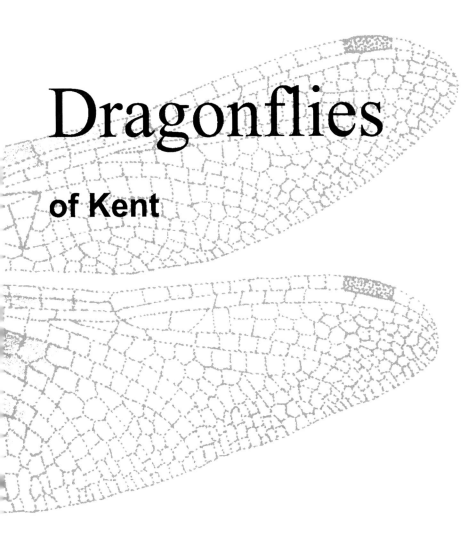

Dragonflies
of Kent

An account of their biology, history and distribution by

John and Gill Brook

Published by the Kent Field Club

Published by Kent Field Club, the Natural History Society of Kent

The aims of the Kent Field Club are to promote an increased interest in natural history, to study and record the distribution and ecology of fauna and flora, and to promote nature conservation in association with the relevant organisations within the County of Kent.

First published 2001
Revised edition 2009

Kent Field Club wishes to acknowledge the valuable assistance given by the Kent and Medway Biological Records Centre in the production of this book.

ISBN 978-0-9561926-1-5

Printed in Great Britain by the
MPG Books Group, Bodmin and King's Lynn

Cover photograph: Ruddy Darter, *Sympetrum sanguineum* (Steve Smith)

Contents

Foreword

Dragonflies are probably our most beautiful insects; they are colourful, elegant, entertaining and fascinating. The true dragonflies, or anisopterans, are powerful fliers, manoeuvring skilfully and appearing almost to enjoy their aerobatic expertise. The damselflies, or zygopterans, are equally enthralling but they operate on a smaller scale and yield their secrets better to more detailed examination.

Dragonflies can easily be observed using close-focusing binoculars and it is usually best to sit down, wait and watch the spectacle that follows. Dragonflies are good indicators of habitat type; for example they can be found in ponds, lakes, boggy pools, streams, rivers and dykes in lowland marshes, and each type of habitat has its characteristic suite of species. They are also excellent indicators of quality of habitat, they are sensitive to pollution and they have responded detectably to climatic changes, even in the last few decades.

John has been interested in natural history from his youth. His enthusiasm was sparked off when he was a member of the Junior Naturalists' Club at Maidstone Museum in the 1950s; this provided an exellent foundation for further study and he learned the elements of taxidermy and entomology there from local professional Eric Philp. John has worked for 35 years on the estate at Leeds Castle while Gill started work as a junior teacher. They both became interested in dragonflies in 1985 and they have visited many areas in Britain following their hobby. They have, however, concentrated their efforts on dragonfly faunistics in Kent and they have added a considerable amount to the knowledge of distribution of the species and their habitats in the county. Their efforts have been particularly effective, especially since 1988, as they have concentrated their work on the exuviae, or cast skins, which are left behind after emergence. These provide definite proof of breeding, whereas a sight record of an adult does not. John and Gill have put a lot of effort into identifying the exuviae of the British species and they have prepared field keys to help others to aquire the same skills.

John and Gill produced the first edition of this work in 2001 and it was a superb attempt to put together all the information that has been gathered on dragonflies in the county. Distribution maps were included and these included information on whether successful breeding had taken place, based upon the evidence of the finding of exuviae. The beautiful drawings of adults and larvae were all made by Gill.

Kent is a rich county for dragonflies with much valuable habitat and it is also well placed geographically to receive any species which are expanding their ranges from continental Europe. Now that more than five years have passed, and a lot more information has come to light, it is time to update the impressive first edition. I would like to recommend this new book to all Kentish naturalists who wish to learn more about this fascinating order of insects and I hope that many will enjoy helping the authors gain more data and monitor the changes of distribution that are taking place.

Graham Vick
Crossfields, Little London, Tadley, Hampshire

Graham's early interest in dragonflies was encouraged by Eric Philp of Maidstone Museum in the early 1970s. He is a founder member and past Vice President of the British Dragonfly Society and he is the UK representative for Odonatologica, the journal of the Societas Internationalis Odonatologica. He has published papers on the dragonflies of Cameroon and Nepal.

Introduction

Shortly after completion of the first *Dragonflies of Kent*, which was published in 2001, we decided that an update should be considered after a further five years of recording Kent's Odonata. This was envisaged as just a few pages of text and maps to be included in the Kent Field Club's annual Bulletin for 2006. However, in 2005 we were asked by Steve Smith of the Kent and Medway Biological Records Centre to prepare a revised edition of the *Dragonflies of Kent*. Much of the original text has been retained with alterations where necessary and the maps have been updated to include all the records received from 1980 to 2008. A new section on exuviae has also been added with a simplified identification key.

Dragonflies were recorded from 821 (71%) of the 1158 tetrads covered by this atlas. Breeding activity was observed from 365 tetrads, with exuviae being collected from 240 of these. However 'up-to-date' the information might be it will of course over time become 'out-of-date' because nature is in a constant state of change. Populations come and go, and fluctuate in numbers as a result of habitat and climate changes. An atlas of the 1940s would have described the Migrant Hawker, *Aeshna mixta* Latreille as an uncommon migrant but today it is now a widespread and common resident species over much of England. And what of an atlas of the future? It might include the Red-veined Darter, *Sympetrum fonscolombii* (Sélys-Longchamps) as a common resident species. Hopefully, it will not record any species as being lost to the county.

We cannot just discuss records and distribution of dragonflies without giving some thought to some of the factors that influence the dragonfly fauna of the county such as Kent's position at the southeastern corner of the British Isles. The temperature is often higher than average and the rainfall moderate, averaging 710mm (28 inches) per annum. This figure is the average rainfall for the years 1993-2005 as recorded at Leeds Castle for the Environment Agency and Meteorological Office. Rainfall measurements will of course vary across the county as other rainfall recorders will confirm. Approximately two-thirds of the county boundary is coastal and so the sea has an influence over the climate. Kent's proximity to the continent is important also, as many migrant species make landfall here after a relatively short flight across the English Channel, especially if aided by a good southeasterly wind. One example mentioned in the Species Accounts is that of *Sympetrum fonscolombii* which has successfully bred in the county for a number of years. A similar and more successful example first recorded in Britain in 1999, is that of the Small Red-eyed Damselfly, *Erythromma viridulum* (Charpentier) which is now widespread in Kent and breeding successfully. Other continental species are also expanding their range just across the Channel, increasing the chance of these species making landfall in Kent.

Contributors and acknowledgements

The following is a list of all those individuals and organisations who contributed records for this project. Without their records, this publication would not exist. The authors would like to thank them all for their time and effort. Every single record of a species, rare or common, is essential to mapping the status of our Kent dragonflies.

List of recorders

Ken Adams, Mary Adams, PG Akers, J Aldis, Jean Aley, GW Allen, R Allen, Martin Allison, D Andrews, The Ash Partnership, Linda Ashton, W Attridge, Gail Austen-Price, John Badmin, Mary Barnard, Anne Barrett, D Batchelor, I Beains, Ian Beavis, Dan Bennett, C Bindon, Chris Birch, C Bladon, Ishpi Blatchley, Fred Booth, Chris Borrow, PR Brash, J Bratten, Gill Brook, John Brook, K Brook-Child, Bryan Bullen, D Carder, Centre For Ecology& Hydrology, Steve Cham, D Chambers, Ruth Childs, Gareth Christian, Tim Christian, Yvette Churchill, P Clarkson Webb, P Claxton, Dan & Charmian Clay, DK Clements, Giovanna Clements, Laurence Clemons, D Coast, R Coles, Jackie Collins, Jerry Collins, Wendy Collins, Hannah Cook, I Cook, Ian Corbet, RP Cordero, J Howard Cox, M Cox, Andrew Craven, Mark Cwynarski, Simon Davey, GAN Davies, Kevin Davis, O Davis, M Davison, Nick Delaney, C Dell, Selwyn Dennis, Nicholas Donnithorne, Mike Dooley, Stuart Dove, Susan Dove, T Dove, M Drake, Jenny Dunn, Chris Dyson, David Eade, Mike Easterbrook, Elham Parish Survey Team, M Ellison, WA Ely, M Enfield, the Environment Agency, H Eve, Alex Ewing, John Feltwell, Glynis Fenn, A Ferguson, Ian Ferguson, Ann Fisher, Bob Fisher, Lynne Flower, Pete Flower, Irene Folliot, Tony Fonseca, Janet Forbes, Alan Ford, PJ Forrest, A Foster, Claire Francis, AE Fray, Deryk Frazer, Kathy Friend, Teresa Frost, Tim Frost, Martyn Gest, EK Goldie-Smith, B Gomes, Mike Gould, DW Grant, A Gray, Phil Grilli-Chantler, S Grove, Alan Hall, Jean Hallwood, Emily Harrington, G Hawgood, B Hawkes, Cliff Hayward, Grant Hazlehurst, Sue Heath, Graham Hemington, A Henderson, Tim Hill, Adrian Hine, Peter Hodge, Ian Hodgson, B Holcombe, A Hold, Vernon Hucks, MP Hunley, N Jarman, M Jennings, Lorinda Jewsbury, A Johnson, Kate Kellett, Kent Wildlife Trust, J Kesby, A Keston, Steve Kirk, Ann Knief, Jacky Langton, Stephen Lemon, DF Lloyd, Lucy Love, Maidstone Museum and Bentlif Art Gallery, R Mann, Lee Manning, Ben Martin, J Martin, M May, J McLaren, Mike Meakins, AV Measday, S Melville, Daphne Mills, BD Moreton, C Morris, N Morris, R Morris, Pat Moss, Richard Moyse, Lys Muirhead, T Mullender, Natalie Muller, MP Musley, Ron Nash, T Neuburger, Wendy Nevard, Martin Newcombe, D Newman, Pete Norwood, Heather Oehl, F O'Hare, Nick Onslow, A Ottley, Mary Page, D Palmer, Margaret Palmer, Alan Parker, David Penney, S Perry, Barry Phillips, Eric Philp, Joyce Pitt, SJ Poole, Mark Pritchard, John Puckett, P Raine, J Richardson, S Ridge, S Ridgen, B Rolleston, V Rook, W Roukin, Lyndsey Rule, J Russell, Alison Ruyter, David Sewell, Mike Sharpe, Judith Shorter, Heather Silk, Sue Skipp, C Slack, Phil Smith, Steve Smith, P Sokoloff, Ann Southern, S Springate, E Still, EL Still, Karen Stone, AE Stubbs, Janet Stubbs, Caroline Sutton, Sue Swales, Don Tagg, J Taplin, Jill Tardivel, Nick Tardivel, Don Taylor, P Thomas, Pat Thompson, B Turner, Derek Tutt, John Tyler, J Varley, B Wain, C Wain, Annie Waite, David Walker, M Walter, Martin Warnes, S Warry, Liz Watkins, Malcolm Watling, John Webley, Liz Weston, W Whitaker, RLP White, Phil Willcocks, J Williams, S Williams, Margaret Willis, Ann Wilson, K Wilson, Shelagh Wilson, David Wiseman, George Witt, Tony Witts, Geoff Woolley, and B Wright. Regional records are also from the Biological Records Centre at Monks Wood, and Alan Hold (former South-East Regional Recorder for the British Dragonfly Society).

Thanks also to all our friends and fellow naturalists in the Kent Field Club for their encouragement and help in all matters of natural history, especially to Eric Philp who encouraged our interest in dragonflies and led us to join the British Dragonfly Society (BDS). Special thanks to Hannah Cook, Steve Smith, Teresa Frost, Ruth Childs, Lee Manning and Irène Folliot of the Kent and Medway Biological Records Centre (KMBRC) for their patient and invaluable help in the preparation of this book; to Richard Moyse of Kent Field Club for preparation of maps and lay-out; to Nick Donnithorne (Surrey Recorder for the BDS) for researching and providing historical records and for his enthusiastic encouragement; Bill Wain (BDS) for proof reading; Graham Vick (BDS) for his invaluable help with identification of exuviae; Peter Allen (BDS) for providing his slides for the illustrations of the Red-veined Darter *Sympetrum fonscolombii* and Globe Skimmer *Pantala flavescens*; and Andrew Chamberlain (BDS) for use of a photograph for drawing Vagrant Darter *Sympetrum vulgatum*. Further thanks must go to Hania Berdys (www.gardensafari.net), Bryan Bullen, Danny Chapman, Nick Donnithorne, Malcolm Farrow, Greg Hitchcock, Lee Manning, Steve Smith, Mike Thurner, Richard Moyse and Kent Wildlife Trust for supply of photographs. Where not otherwise credited, photographs are by Gill Brook. All species illustrations were drawn by Gill Brook.

Maps were produced using DMAP by Dr Alan Morton.

The following are acknowledged for kindly allowing access to private land:

Mr. & Mrs. Bainbridge, Bedgebury School, Mrs. Bloemers, Mr. Boatridge, Robert Brett & Sons Ltd., Chart Hills Golf Club, Mrs. Cornish, Mrs. G. DeVilleneuve, Mr. S. Dove, Mr. & Mrs. T. Dove, Mr. & Mrs. Farmer, Mrs. D. Franks, Mr. S. P. Gent, Mr. & Mrs. Harman, Mr. J. Hawkes, Mrs. E. Hussey, Mr. D. Johnson, Mr. & Mrs. Q. Johnson, The Viscount and Viscountess Monckton of Brenchley, P. Morris, The National Trust (Sissinghurst Castle), Mrs. Parish, Mr. Pullen, Mr. Pym, Mrs. Richardson, The Ridge Golf Club, Sheerness Driving Range, Sheerness Golf Club, Mr Stannard, Vesper Hawk Farm, Mr. Wallis, Mr. & Mrs. P. Ward, The Weald of Kent Golf Club, Mrs. Wraight, Mr. & Mrs. Young and those who expressed a wish to remain anonymous.

Checklist of Odonata found in Kent

The sequence and scientific nomenclature follows Merritt, Moore and Eversham (1996).

≠ Accidental introduction to Kent

Migrant species to Kent which *have not* bred in the county

* Migrant species to Kent which *have* bred in the county

+ Migrant species to Kent which *have attempted* to breed

X Species formerly recorded in Kent, but not recorded during the period of the survey

Ovipositing by Large Red Damselfly (Nick Donnithorne)

Suborder ZYGOPTERA

Family CALOPTERYGIDAE
Calopteryx virgo (Linnaeus, 1758) Beautiful Demoiselle

Calopteryx splendens (Harris, 1782) Banded Demoiselle

Family LESTIDAE
Lestes viridis (Vander Linden, 1825) * Willow Emerald Damselfly

Lestes barbarus (Fabricius, 1798) + Southern Emerald Damselfly

Lestes sponsa (Hansemann, 1823) Emerald Damselfly

Lestes dryas Kirby, 1890 Scarce Emerald Damselfly

Family PLATYCNEMIDIDAE
Platycnemis pennipes (Pallas, 1771) White-legged Damselfly

Family COENAGRIONIDAE
Pyrrhosoma nymphula (Sulzer, 1776) Large Red Damselfly

Erythromma najas (Hansemann, 1823) Red-eyed Damselfly

Erythromma viridulum (Charpentier, 1840) Small Red-eyed Damselfly

Coenagrion puella (Linnaeus, 1738) Azure Damselfly

Coenagrion pulchellum (Vander Linden, 1825) Variable Damselfly

Enallagma cyathigerum (Charpentier, 1840) Common Blue Damselfly

Ischnura elegans (Vander Linden, 1820) Blue-tailed Damselfly

Suborder ANISOPTERA

Family AESHNIDAE

Aeshna juncea (Linnaeus, 1758) ˣ Common Hawker

Aeshna mixta Latreille, 1805 Migrant Hawker

Aeshna affinis Vander Linden, 1823 ˣ Southern Migrant Hawker

Aeshna cyanea (Müller, 1764) Southern Hawker

Aeshna grandis (Linnaeus, 1738) Brown Hawker

Anax imperator Leach, 1815 Emperor Dragonfly

Anax parthenope Sélys-Longchamps, 1840 # + Lesser Emperor

Hemianax ephippiger (Burmeister, 1839) ˣ Vagrant Emperor

Brachytron pratense (Müller, 1764) Hairy Dragonfly

Family GOMPHIDAE

Gomphus vulgatissimus (Linnaeus, 1758) ˣ Common Club-tail

Family CORDULEGASTRIDAE

Cordulegaster boltonii (Donovan, 1807) Golden-ringed Dragonfly

Family CORDULIIDAE

Cordulia aenea (Linnaeus, 1758) Downy Emerald

Somatochlora metallica (Vander Linden, 1825) Brilliant Emerald

Family LIBELLULIDAE

Libellula quadrimaculata Linnaeus, 1758 Four-spotted Chaser

Libellula fulva Müller, 1764 Scarce Chaser

Libellula depressa Linnaeus, 1758 Broad-bodied Chaser

Orthetrum cancellatum (Linnaeus, 1758) Black-tailed Skimmer

Orthetrum coerulescens (Fabricius, 1798) Keeled Skimmer

Sympetrum striolatum (Charpentier, 1840) Common Darter

Sympetrum vulgatum (Linnaeus, 1758) # Vagrant Darter

Sympetrum fonscolombii (Sélys-Longchamps, 1840) ˟ Red-veined Darter

Sympetrum flaveolum (Linnaeus, 1758) # Yellow-winged Darter

Sympetrum sanguineum (Müller, 1764) Ruddy Darter

Sympetrum danae (Sulzer, 1776) # Black Darter

Leucorrhinia pectoralis (Charpentier, 1825) ≠ Large White-faced Darter

Pantala flavescens (Fabricius, 1798) ≠ Globe Skimmer

Freshly emerged Black-tailed Skimmer (Nick Donnithorne)

A brief history of Odonata recording in Kent

Common Club-tail - no longer found in Kent (Brian Bullen)

Scarce Chaser - recorded before 1845 "near Herne Bay" (Lee Manning)

"There are few English counties which had a more interesting insect fauna than Kent" according to William Page, editor of 'The Victoria History of the County of Kent', published in 1908. He continued by saying that "dragonflies seem to have received less attention than has been given to this order in many other counties".

In the 1845 publication 'British Libellulinae or Dragonflies', the author, W. F. Evans, mentioned several species for Kent, giving some locations as a specific piece of woodland or stream. For example, he reported that the Beautiful Demoiselle, *Agrion (Calopteryx) virgo* was abundant at a small stream near Blean Wood. In Kent today this species is only recorded from the west and south-west of the county near to Sussex, a long way from Blean Wood, which lies north of Canterbury.

Another interesting record was of the Scarce Chaser, *Libellula fulva* for which he just gave "near Herne Bay" as the location, which is not very far from Westbere Marshes where *L. fulva* is now found and known to breed. Other records, such as those for the Broad-bodied Chaser, *L. depressa* and the Southern Hawker, *Aeshna cyanea*, are just given as "common in Kent" and "abundant in Kent". He also mentioned species as being found "round London" and "neighbourhood of London and Metropolitan District" which are too vague to be claimed for Kent.

Some years later, in 1890, W. H. Bath produced 'An Illustrated Handbook of British Dragonflies'. He listed fewer species for Kent than Evans and generally only gave locations by town names. An interesting record was that of the Red-veined Darter, *Sympetrum fonscolombii* at Deal, not far from one present location at Samphire Hoe. This species was no doubt a migrant from Europe just as it is today, although it is now establishing itself as a resident in Kent. It is also interesting to note that he recorded the Common Club-tail, *Gomphus vulgatissimus* from Dartford, a species not now found in Kent.

At the turn of the century, the picture of the status of our dragonflies became clearer with the publication of 'British Dragonflies' by W. J. Lucas in 1900. Among the records is one for the Variable Damselfly, *Coenagrion pulchellum* from Deal and Canterbury, approximately the same distribution as for today. A more unusual record was for the European species the Large White-faced Darter, *Leucorrhinia pectoralis* (Charpentier) near Sheerness in June 1859 and attributed to Bath. However, it is not known if this dragonfly made landfall in Kent by itself. It is believed that it may have been taken on board a fishing boat (McLachlan, 1884). Later, in January 1860, it was exhibited at a meeting of the Entomological Society of London. There are no other records for this species in Britain except for an unconfirmed sighting near Chartham in May 2000. The nearest locations are in Belgium and in the Netherlands, Germany, southern Scandinavia, southern Europe and eastwards to Siberia. There are also a few locations in north east France. Lucas gave the source for all his records and reported a better coverage of the county than previous authors had. However, most of his work appears to have been from Surrey, the county in which he lived.

The locations for records were still given by town names or more general terms and this seems to have been the case at least up to and including the publication of the aforementioned 'Victoria History of the County of Kent' in 1908. The editor, W. Page listed records for twenty species in Kent, along with sources and locations. The bulk of locations were from coastal areas in the east of the county from Folkestone to Margate. Inland locations included Canterbury, Appledore and Chattenden, and also Lee and Deptford, now part of London. Among his records is one for the Keeled Skimmer, *Orthetrum coerulescens* at Chattenden attributed to H. J. Turner.

Variable Damselfly - pre-1910 records show a similar distribution to today

Freshly emerged Emperor Dragonfly

Red-eyed Damselfly (Nick Donnithorne)

In recent years Hothfield Common, near Ashford, was thought to be the only known site in Kent for this species until two adult specimens were reported in 2003 from the Kemsley area (see *O. coerulescens* in Species Accounts). He also reported that he had "no recent records of the Emperor Dragonfly, *Anax imperator*" but gave Evans and McLachlan as historical sources for this species in Kent. Today *A. imperator* is widespread in the county.

As a schoolboy and later on leave from National Service, Norman W. Moore studied dragonflies on the Kent and Sussex border. Between the years 1938 and 1947, he covered the Rother Levels and Isle of Oxney in Kent, as well as an area between Rye and Lewes in Sussex. He was responsible for discovering populations of the Scarce Emerald Damselfly, *Lestes dryas* and the Variable Damselfly, *Coenagrion pulchellum* in the well vegetated-ditches of the floodplain, particularly around Wittersham and Newenden. Unfortunately, there have been no further records of these species at these locations. Dr. Moore is one of the foremost experts on the dragonflies of Britain today. All his data were given to the Biological Records Centre (BRC) at the Institute of Terrestrial Ecology, Monkswood, Cambridgeshire.

'The Dragonflies of the British Isles' second edition by Cynthia Longfield was published in 1949 (Longfield, 1949a). Records are listed in very general terms such as "reported in Kent" for the Common Hawker, *Aeshna juncea,* a species not presently recorded in the county, though there are occasionally unconfirmed verbal reports. *Coenagrion pulchellum* was thought to have "died out", the Red-eyed Damselfly, *Erythromma najas* "recorded in the past" and *Orthetrum coerulescens* "no recent records". These last three species are to be found in Kent today.

In 1955, the Kent Field Club (KFC) was formed to study and record all groups of the county's wildlife. Records, observations and studies are published in the Club's annual Bulletins and the periodical Transactions. Since the Club's formation until the present day, there has been a wealth of expertise in the membership. Among these was Dr. G. A. N. Davis who was the Club dragonfly recorder from 1959 to 1970.

In 1983 a national body, the British Dragonfly Society (BDS), was formed, which has done much to promote interest in these insects. Members of both the Kent Field Club and British Dragonfly Society have kept records that are included in this atlas.

The Kent Biological Archive and Records Centre formerly based in Maidstone Museum was begun in about 1971 by Eric Philp who was the Keeper of Natural History there from 1956-1993. He was keeping records before taking up this post, but as his records and those sent to him by other recorders increased, he saw the need for a central collecting and collating system. Among the records are many for dragonflies and these are also included in this publication.

The current Keeper of Natural History, Dr Ed Jarzembowski, has a great passion for fossil insects and he informs us that the oldest known dragonfly from Kent is a fragmentary fossil collected from the Wadhurst Clay Brickworks, Tonbridge, reported by Crowson (1946). It is either a true dragonfly (Anisoptera) or a species of Anisozygoptera (a mainly extinct Suborder) with only two species surviving in the world today. The specimen dates from the Cretaceous Period, 135 million years ago.

What then of the future? Recording the dragonflies of Kent will continue to be important so that changes in populations and habitats can be noted and monitored. It is hoped that this atlas will encourage dragonfly enthusiasts, whether amateur or professional, members of societies or not, to continue their recordings and to pass them on to the authors or the Kent and Medway Biological Records Centre at Brogdale Farm, Brogdale Road, Faversham (see Useful addresses, page 121). The authors would also like to encourage recorders to collect and identify the exuviae. Not only do they provide proof of breeding through to adulthood but they also provide a record of species when adults are not seen. Unfortunately, many dragonfly enthusiasts take little interest in this rewarding and challenging aspect of dragonfly recording.

About dragonflies

Dragonflies are winged insects with a typical three-part body made up of head, thorax and abdomen. The thorax has three pairs of legs and two pairs of wings attached. All British species require aquatic habitats for breeding and all are predatory feeders in both the larval and adult stages.

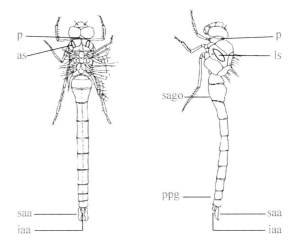

Figure 1: A typical male Anisopteran dragonfly; **as**, antehumeral stripe; **iaa**, inferior anal appendage; **ls**, lateral stripe; **ppg**, position of primary genitalia; **p**, prothorax; **saa**, superior anal appendage; **s(a)go**, secondary (accessory) genital opening.

Dragonflies fall into two convenient divisions, or sub-orders, which are soon easily recognised. Those called dragonflies, the anisopterans, are generally larger or more robust than the damselflies, or zygopterans, which are more delicate in build and in flight. The anisopterans have large, almost 'wrap-around', eyes that meet at the top of the head (fig. 1), their fore wings are different in shape to their hind wings (figs. 2-3) and they rest with their wings held out straight at right angles to the body. Some species, especially the libellulids, may sometimes rest with wings held forward and angled downwards. The Common Club-tail, *Gomphus vulgatissimus* Linnaeus, not found in Kent, differs from other anisopterans in having widely spaced eyes. Zygopterans have widely separated almost spherical, eyes. The fore and hind wings are the same shape (figs. 4-5) and are usually held closed along the abdomen when at rest. However, species of the genus *Lestes* usually hold their wings half open at about 45° to the body.

All dragonflies are accomplished hunters, catching most of their prey on the wing. Almost any insect that flies is fair game including true flies, mayflies, wasps, butterflies and even other dragonflies. The authors have observed the Golden-ringed Dragonfly, *Cordulegaster boltonii* (Donovan) eating a common wasp, *Vespula vulgaris* (Linnaeus) and the Southern Hawker, *Aeshna cyanea* (Müller) eating the soft abdomen of a White Admiral butterfly, *Ladoga camilla* Linnaeus, discarding the rest of the insect.

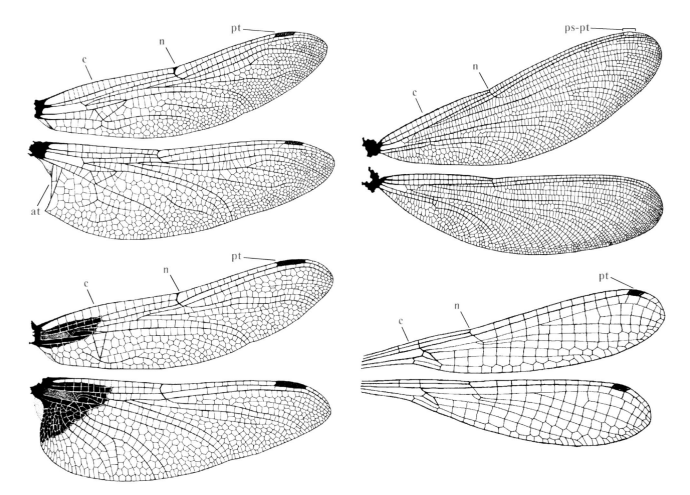

Figures 2-3: Typical Anisopteran wings
Fig. 2 (above) *Aeshna cyanea*, male
Fig. 3 (below) *Libellula depressa*, female
at - anal triangle; **c** - costa; **n** - node; **pt** - pterostigma.

Figures 4-5: Typical Zygopteran wings
Fig. 4 (above) *Calopteryx splendens*, female
Fig. 5 (below) *Coenagrion puella*, female
c - costa; **n** - node; **ps-pt** - pseudo-pterostigma; **pt** - pterostigma.

Damselflies usually catch prey, such as aphids, that are resting among the foliage. Their success as hunters is largely due to their good eyesight and exceptional agility in flight. Their eyes are compound in structure with up to 28,000 individual lenses, or facets per eye in the larger species enabling them to recognise and follow prey and to recognise territorial boundaries. Their sight appears quite exceptional. Unlike many other four-winged insects, dragonfly wings have no devices linking the fore and hind wings together, and the flight muscles are directly attached to the wing roots not to the exoskeleton. Each of the four wings can be operated individually and the pitch can be varied. This gives them great manoeuvrability and enables them to make sudden changes in direction. They are the master flyers of the insect world.

Some dragonflies are territorial in behaviour, especially the larger species. They can often be seen patrolling a length of waterway, turning at exactly the same point each time. Rivals of the same species are vigorously chased away. Some species, especially libellulids, will continually return to the same perch, rather in the manner of the Spotted Flycatcher, *Muscicapa striata* (Pallas). Dragonflies are often encountered far from water, especially in woodland rides and hedgerows where insect prey is abundant, but they return to water bodies to search for mates and for breeding.

The mating style is unique among insects, the male having primary genitalia near the abdomen tip, and secondary genitalia under abdominal segment 2. In flight prior to mating, the male anisopteran bends his abdomen double so that he can transfer sperm from the primary to the secondary genitalia. Once this priming has taken place he grasps the female, in flight, by the thorax and brings his abdomen forward so that he can grasp her head with the anal appendages at the tip of his abdomen. He then releases his leg hold on her thorax and the pair fly *in tandem*. The zygopteran male grasps the thorax of the female with his legs while she is at rest and then primes his secondary genitalia. Grasping her prothorax by means of his anal appendages (fig. 6), the pair then fly *in tandem* for a while. When ready for copulation the female dragonfly brings her abdomen forward and under to come in contact with the secondary genitalia of the male. This may take place in flight or at rest and this position is known as the 'mating wheel' (fig. 7). Once copulation is complete, the female is ready to begin ovipositing and this may be while still *in tandem* or after the pair have separated (figs 8 and 9). In some species, the separated male may stay in attendance near the female to prevent rival males attempting to mate with her. Species that scatter their eggs onto the water surface have spherical eggs and those that inject them into plant material have elongated eggs (Hammond, 1983). More details of ovipositing for particular species can be found in the species accounts section. The eggs hatch within two to four weeks unless they enter diapause in which case they pass the winter as eggs and then hatch in the following spring.

Like many other insect orders, the longest part of the life span is spent in the larval stage. Larval development can take from less than one year or up to four years before becoming adults, depending on the species and conditions such as temperature and food abundance. Larvae will often wait for prey to pass by and grab it with a lightning movement of their labium or 'mask'. At other times, they will stalk their intended prey with agonisingly slow movements that will place them in position to use their labium. Insect larvae, including those of other dragonflies, tadpoles and small fish are just some prey species taken. Frogs, fish, water beetles and newts are some of the species, which in turn prey on dragonfly larvae. Some species of larvae live in leaf litter or mud on the bottom of the

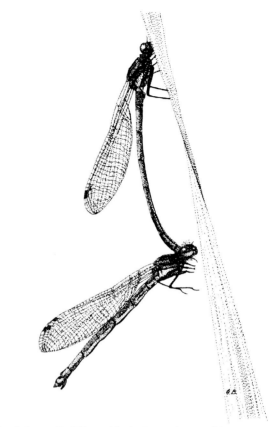

Fig. 6: Large Red Damselfly in the tandem position

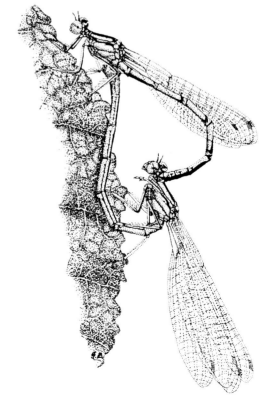

Fig. 7: White-legged Damselfly in the 'mating wheel'

Fig. 8: Emperor Dragonfly ovipositing in submerged plant tissue

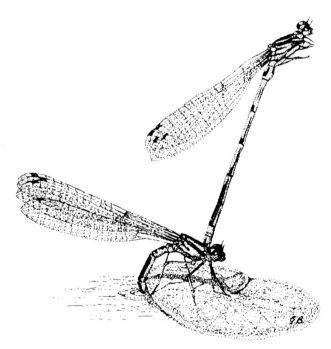

Fig. 9: Azure Damselfly ovipositing with the male in the sentinel position.

water body, while others live among aquatic vegetation.

Anisopteran larvae are robust and have internal gills, and may have lateral or dorsal spines on their abdominal segments. They crawl on the bottom of the water body, or move through the water at great speed by 'jet propulsion'. This is achieved by rapidly passing water out of the anus, thus moving them forward by this thrust. Zygopteran larvae are slim and streamlined in shape. Their gills are external leaf-like lamellae attached in threes to the tip of the abdomen. They crawl on the bottom or swim through the water with bending and flexing movements similar to the swimming movement of sharks. After passing through about 10-15 instars, the larva is ready for adult emergence. It seeks shallower water and a suitable support, such as a plant stem, on which to cling while emergence takes place. It may pause for a while as it leaves the water to take in extra oxygen from the air, which it needs to help it through this important and vulnerable stage. With the legs firmly grasping the chosen support, the thorax and head burst through the larval skin and then the legs too are withdrawn. A resting stage now occurs with anisopterans hanging down (fig. 10), and zygopterans resting in a vertical position. During this time, the newly exposed legs harden so that they can grip the support firmly while the abdomen is finally withdrawn. The abdomen then gradually extends to full length, about twice that of the larval abdomen, during which time it pumps fluid into the wings, gradually inflating them to full size (fig. 11). A period of drying out and hardening then follows and body weight is reduced all in preparation for its maiden flight. The empty larval skins or exuviae remain clinging to the support (fig. 12) and may be collected later by the dragonfly observer for identification. This first flight will take it to a safe area such as a woodland ride or hedgerow where it can feed, gain its full coloration and become sexually mature. This may take from a few days to a week or two depending on the species. It is then ready for mating and the life cycle can begin again.

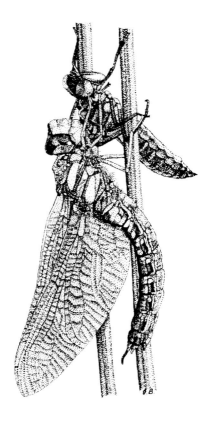

Fig. 10 (above, far left): An emerging *Libellula depressa* in the anisopteran resting stage

Fig. 11 (above left): A fully emerged *Aeshna* sp. still clinging to its exuviae

Fig. 12 (above): Exuviae of *Libellula fulva* still attached to the support.

Once the larvae leave the water and begin adult life they are vulnerable to predation by birds and small mammals. Sparrows, robins and blackbirds seemed to know when the dragonflies were emerging from the author's pond and would gather at the edge to pick off the larvae, pre-flights and tenerals as they took their maiden flight. The most well known avian predator is the Hobby. Other predators include spiders, when dragonflies become entangled in their webs. It is also not uncommon to find small dragonflies, especially damselflies, caught on the sticky leaves of the insectivorous sundew plant (*Drosera* spp.). The authors were surprised to find on one occasion remains of three or four Black Darters, *Sympetrum danae* (Sulzer) protruding from small holes in bare ground near a pond. These holes were thought to have been made by an unknown species of predatory insect. Perhaps the inhabitants of the holes took the opportunity to grab the dragonflies while they were at rest on the ground by the holes.

Should the dragonflies escape all these dangers and others such as parasites and disease, they may reach an adult age of about 30-60 days. No British species has been known to hibernate through the winter even though some species such as the Common Darter, *Sympetrum striolatum* (Charpentier) are often still on the wing in late November or early December if weather conditions are suitable.

The authors have often been asked "What use are dragonflies; what good do they do?" They are useful pollution indicators. Although a couple of species will tolerate mildly polluted water, by far the majority will only breed in and inhabit good quality water bodies. Dragonflies are also useful in the control of pest species such as aphids, mosquitoes and midges, which form a significant part of their diet. The other reason is purely aesthetic, for their amazing agility in flight and their colourful beauty are obvious. The natural world would be all the poorer should they disappear.

Dragonfly habitats

Kent is a county of varied landscapes and topography due to the underlying geology. Rock and soil types occur in a series of parallel and sometimes overlapping bands that run approximately east-west across the county.

From north to south, they are London Clay, Eocene Sands, Clay-with-flints, Chalk, Gault Clay, Folkestone Beds, Hythe Beds, Wealden Clay and Tunbridge Wells Sands. In the far south of the county is Europe's largest expanse of shingle. These bands produce the varied landscapes that in turn, with man's use of the countryside, dictate the types of habitats available to dragonflies. For example, gravel extraction has led to many deep-water gravel-pit lakes and drainage of the marshes has given us excellent dyke habitats.

An amenity lake associated with new housing

After quarrying for gravel or other minerals has ceased many pits become flooded resulting in large deepwater lakes often with a narrow shallow shelf around the shoreline. Many gravel-pit lakes, as at New Hythe and Larkfield, have been stocked with fish and are typically frequented by the Black-tailed Skimmer, *Orthetrum cancellatum*. Wooden platforms that have been erected around the lakes provide ideal places from which to observe dragonflies and from which to search for exuviae. In recent years, many fishing lakes have been constructed on farmland offering a similar habitat to that of gravel-pit lakes. Other lake habitats include ornamental lakes usually with a shallower profile that allows more aquatic vegetation to become established as at Marshall's Lake in Bedgebury Pinetum. It is pleasing to find that new housing developments frequently include a pond or lake for local residents to enjoy and quickly become populated with dragonflies.

At the southern extremity of Romney Marsh is Denge Marsh, an extensive area of shingle dunes.

Newly emerged male Black-tailed Skimmer, *Orthetrum cancellatum* (Nick Donnithorne)

Royal Military Canal, near Appledore (RI Moyse)

This is a huge SSSI by virtue of its unique flora and fauna and its nearness to continental Europe, which makes it an ideal landing place for many migrant birds and insects. The Lesser Emperor, *Anax parthenope* is regularly seen at some of the lakes and the Red-veined Darter, *Sympetrum fonscolombii* has been known to breed here. There are many gravel-pit lakes of various sizes in the part of the site that is managed by the RSPB and another part is home to the Dungeness

Bird Observatory. Most of the SSSI is open to public access.

An Ordnance Survey map of the Weald of Kent will show an area that is peppered with numerous farm ponds. Some of those shown may no longer exist or may be neglected because of changes in farming needs and practices. However, well maintained species rich farm ponds can still be found and farmers in general have been happy to give permission to survey their ponds for dragonflies. Some woodland ponds have suffered from neglect where woodland management has declined but there are still good examples to be found which are potentially valuable habitats for Kent's two Cordulid species. A growing interest in wildlife and conservation has led to an improvement in some of our village ponds through the efforts of local volunteers and the BTCV pond warden scheme. These ponds and garden ponds are valuable as dragonfly habitats especially on chalk where surface water is scarce.

Kent has extensive marshlands along the Thames estuary and north coast, which are collectively known as the North Kent Marshes. Running across country from Reculver to Sandwich are the Chetney Marshes and Ash Levels, and in the south is the large area known as Romney Marsh that is separated from the Weald by the Royal Military Canal (RMC). For centuries all these marshland areas have been drained by a network of small dykes, typical habitats for the Hairy Dragonfly, *Brachytron pratense*, linked to larger ones that in turn are linked to nearby rivers. The habitats that they provide are as varied as the dykes are; some of those nearest to the coast, as at Cliffe, may be brackish. Although the RMC is large, like a river, its still waters provide a habitat similar in character to that of a large dyke.

Some of the rivers, such as the River Medway, are typical of lowland Britain having slow and cloudy waters with little or no submerged aquatic plant life. Others, such as the Rivers Darent and Great Stour, have clearer waters of moderate flow with some aquatic vegetation. In places the River Darent has a very gravelly bed. Smaller rivers, like the River Beult, may have such plants as duck weed and water lilies, as do some of the best species rich streams. The smallest streams with a faster flow and clear water in or near woodland, such as at the NT property at Sissinghurst, may sometimes be home to the Golden-ringed Dragonfly, *Cordulegaster boltonii* or the Beautiful Demoiselle, *Calopteryx virgo*.

River Beult

White-legged Damselfly, *Platycnemis pennipes*, a species found in the River Beult (Nick Donnithorne)

19

Hothfield Common is one of Kent's last remaining heathland sites with the county's only true acid bog habitat. It is also the only confirmed site in Kent where the Keeled Skimmer, *Orthetrum coerulescens* is known to breed and where it may be seen. There is a selection of typical acid bog flora there, such as Bog Asphodel, Round-leaved Sundew, Heath-spotted Orchid and Lousewort. Place names on the map which follow the Folkestone and Hythe Beds, such as Penenden Heath, Langley Heath, Lenham Heath and Charing Heath, suggest that this habitat may once have been more widespread in Kent.

The rush-filled dune slacks at Sandwich Bay are rare in Kent and are a typical habitat for the Southern Emerald Damselfly, *Lestes barbarus* that arrived in Kent from the continent in 2003. In 2004, it attempted to breed, but unfortunately, the site was flooded by seawater in early 2005.

A pond at Hothfield Common

Introduction to the species accounts and maps

In Kent, a total of 40 species of dragonflies have been recorded. This number includes 27 resident breeding species, 7 migrants, 4 historical species and 2 accidental introductions.

The description of each species is given for the dorsal view of the insect with occasional reference to the side of the thorax where appropriate. Although quite detailed, these descriptions are not fully complete and should be used in conjunction with a good field guide for identification. The same applies to the black and white illustrations. A few field guides are listed in the reference section. As an extra aid to the identification of the blue damselflies of the family Coenagrionidae, an illustration is included (figs. 13-14). This shows the details of the abdominal segments 2, 9 and 10 of the three species found in Kent.

Each account also includes a brief indication of typical habitats, some basic breeding information and a summary of the main areas of distribution in mainland Britain and Kent in particular. A separate note on status is given for the four nationally important species that occur in Kent. The British Dragonfly Society has drawn up a provisional list of regionally important species for each vice county. Those species that occur in VC15 and VC16 are listed under this status. Those species that are migrants or accidental introductions are identified as such in the accounts text. In all other cases, the distribution plotted on the maps will indicate the county status for each species.

The distribution maps show all the records received during the survey period, from 1980-2008. Contributors were asked to submit records with at least the minimum information of species name, date, a four-figure map reference and observers' name. They were also encouraged to give additional information on frequency, habitat and breeding activity, and to search for exuviae as proof of breeding. All maps show the main river and canal

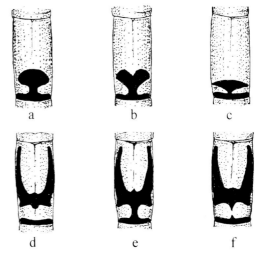

Fig. 13: 2nd abdominal segment of (a) *Enallagma cyathigerum*, typical form; (b) variation of *E. cyathigerum* from Murston Lakes (TQ96H); (c) variation of *E. cyathigerum* from Preston Marsh (TR26K); (d) *Coenagrion puella*, typical form; (e) *C. pulchellum*, typical form; (f) variation of *C. pulchellum*.

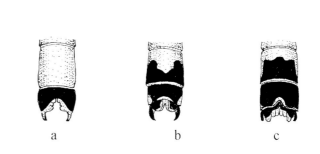

Fig. 14: 9th and 10th abdominal segments of (a) *E. cyathigerum*; (b) *C. Puella*; (c) *C. pulchellum*.

systems and are divided into 10Km squares. The records are plotted in tetrads of which there are 25 in a 10Km square and each tetrad is identified by a letter to indicate its position (fig. 15). Figure 16 shows all tetrads for which records have been received. Some tetrads contain key sites with numerous species and numbers of dragonflies, whereas others contain just casual sightings, and figure 17 shows the number of species which have been recorded from each tetrad. There are 54 tetrads where 15 or more species were recorded, with TQ55I, TQ63X, TQ64B, TQ64I, TQ67V, TQ73B, TQ73G, TQ83E, TR01Y and TR35N having 19 or more species. Some of the species-rich sites are Shorne Wood Country Park (TQ66Z and TQ67V), Holborough Marsh (TQ76B), Furnace Lake (TQ43P), Haysden Country Park (TQ54S and TQ54T), Leeds Castle (TQ85G), Chart Hills Golf Course (TQ84K), Bedgebury Pinetum and Forest (TQ73G), Hothfield Common (TQ94S), Dungeness (TR01U, TR01Y, TR01Z, and TR02Q), Westbere Marshes (TR16V), Ham Fen (TR35H) and Sissinghurst Castle Lakes (TQ83E). As some of the best sites do not have public access, the authors have chosen a selection of those which can be visited which are listed after the species accounts.

Three symbols are used on the distribution maps, a yellow dot for a sighting of a mature adult only, an orange dot for breeding where just mating, egg laying or larvae have been observed, and a red dot for successful breeding where exuviae have been found, thus proving a fully completed life cycle from egg to adulthood. In this publication the word exuviae refers to the rigid, empty cast skins of the larvae from which the adults have emerged and are most commonly found clinging to vegetation usually at or near the water margin.

The land boundary of the distribution maps is a combination of the present day administrative boundary and that of vice-county 16 (West Kent). County boundaries have been occasionally altered adding to the county at one location and taking away from it at another. The ideal solution is to have permanent boundaries to work with and this is why the Watsonian Vice County system is most often used by naturalists. These vice county (V.C.) boundaries are based on old administrative boundaries and are permanent. Kent has two, V.C.16 for West Kent and V.C.15 for East Kent. However, because there are parts of present day Kent not included in V.C.16, the current administrative boundary has also been incorporated. Incidentally, the Kent Red Data Book has also adopted this combination system (Waite, 2000).

Finally, a few words about blank tetrads. If an area of the map is blank, it does not necessarily indicate that dragonflies do not occur there. It may mean that no recording has taken place there or that no dragonflies were seen during a recorder's visit. Many tetrads, on chalk downland for example, are blank because there are few permanent waterbodies except for occasional farm and garden ponds and so it should be no surprise that species are scarce in those areas. Intensively cultivated areas may also lack records. The Isle of Thanet is both chalk and intensively cultivated and this may be why there are so few records for this area.

Fig. 15: The arrangement of letters for the 25 tetrads in a 10 x 10 km square of a standard Ordnance Survey map. The red dot represents TQ85G.

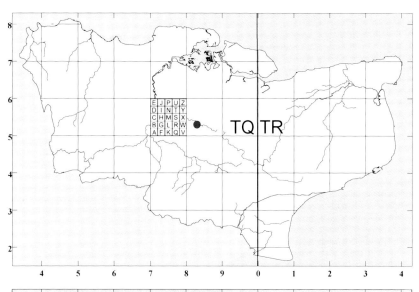

Fig. 16: All tetrads for which records have been received.

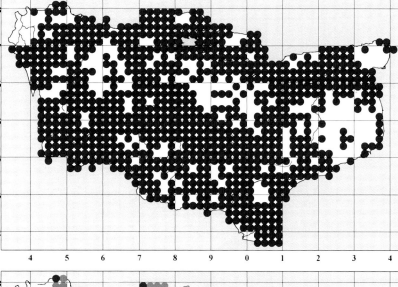

Fig. 17: Total number of species recorded in each tetrad.

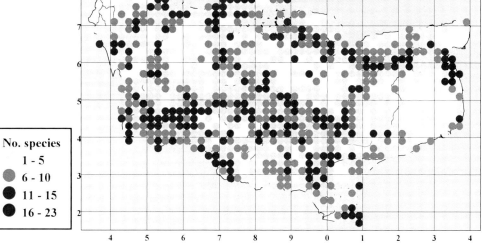

No. species
1 - 5
6 - 10
11 - 15
16 - 23

Beautiful Demoiselle

Calopteryx virgo (Linnaeus)

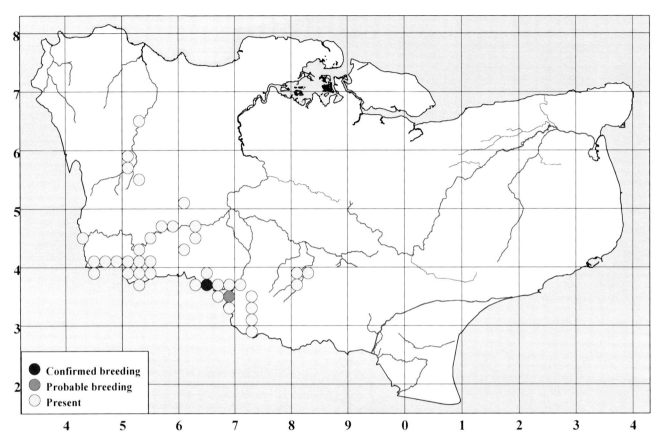

Description

The male of this large damselfly has a metallic blue-green thorax and abdomen, and wings that are almost entirely brown-black. The base and tip of the wings are sometimes paler but the actual amount is variable. The female is metallic green and has brownish wings with white pseudo-pterostigmata. She may have bronze highlights at the tip of her abdomen.

Beautiful Demoiselle, female

Habitat

This species prefers clear, fast-flowing, stony or sandy-bottomed streams and rivers, with some submerged vegetation. It is more tolerant of shade than *C. splendens* and may sometimes be found in woodland and scrub.

Breeding

The male performs a courtship display for the female by frequently flicking his wings half open whilst perched nearby, or whilst in flight. The female oviposits into the tissue of a variety of aquatic plants while the male often stays in close attendance, but she rarely becomes submerged. After about two years the larvae are ready to emerge. Exuviae are found on marginal vegetation close to the water's edge or occasionally a few metres away. They have a distinctive shape with long legs and antennae.

Beautiful Demoiselle, male

Distribution

Calopteryx virgo is mainly found west of a line from Liverpool to Dungeness, with isolated populations in north-west and north-east England, and western Scotland. In Kent, *C. virgo* was said to be locally common (Longfield, 1949a), and also that it was found on the River Darent (Longfield, 1949b). Recent records show them as occuring in the south-west quarter of the county, concentrated mainly to areas near the county boundary. They are also still found along parts of the River Darent. There was an unconfirmed record of this species on a tributary stream of the River Great Stour near Naccolt in 2005.

Status

Regionally important in VC16

Similar species

Calopteryx splendens - the male *C. splendens* rarely has entirely coloured wings.

Flight period in Kent

Mid May to end of August.

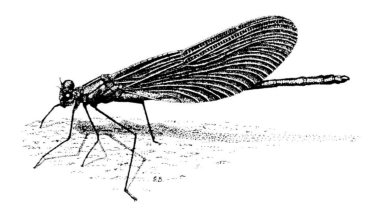

Banded Demoiselle
Calopteryx splendens (Harris)

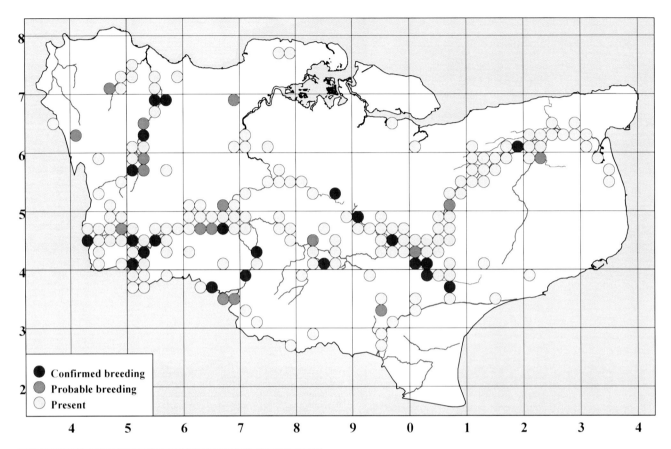

Description

This species is one of the largest of the damselflies. The male has a metallic blue-green thorax and metallic blue abdomen. The wings of the male have a broad, dark, blue-black band giving rise to its English name of 'Banded'. On each side of this band the wing is clear and there are no pterostigmata. The female is dark metallic green with bronze highlights on the thorax and on both ends of the abdomen. Her wings are greenish with white pseudo-pterostigmata. The flight of this genus is very butterfly-like.

Habitat

Calopteryx splendens can be found in slow flowing rivers, streams and canals, with muddy bottoms, running through open countryside, such as meadows and arable fields. There is usually tall emergent and marginal vegetation.

Banded Demoiselle, male (Steve Smith)

26

Breeding

The male performs a courtship display for the female, similar to that of *C. virgo*, by frequently flicking his wings half open while perched nearby, or while in flight. After copulation, with the male often in attendance nearby, the female will oviposit into the tissue of a large variety of aquatic vegetation. She may even climb right down into the water, becoming completely submerged. Larval development usually takes two years. It is recorded that the larvae will sometimes travel a considerable distance from the water, maybe as much as 100 metres, to emerge on shrubs and trees (Brooks, 1997), but the authors have usually found them over the water on rushes, reeds and Sweet grass (*Glyceria* sp.). The exuviae of this genus are a very distinctive shape with very long legs and antennae.

Distribution

It is found in England and Wales, mainly south of a line from the Rivers Humber and Ribble, though a few isolated populations do occur north of this line. In Kent the distribution closely follows the main rivers and their tributary streams. It is often abundant. It is sometimes seen at ponds and lakes. A larva was found by the authors in a pond at Shorne Wood Country Park. On another occasion numerous wings of male *C. splendens* were found on the path by the River Medway, near Hadlow. These were discarded by feeding pied wagtails. An individual *C. splendens* was observed by J. Badmin repeatedly attempting to enter a shop window in Canterbury for over twenty minutes, perhaps seeing its reflection as a rival or mistaking the glass for water.

Similar species

Calopteryx virgo - the colouration of the males' wings is more extensive than that of *C. splendens*. Females of these two species can sometimes be difficult to separate.

Flight period in Kent

Early May to mid September.

Banded Demoiselle, female

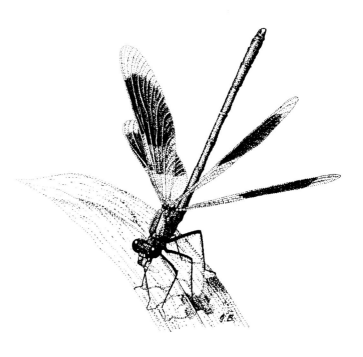

Willow Emerald

Lestes viridis (Vander Linden)

Suborder Zygoptera

Family Lestidae

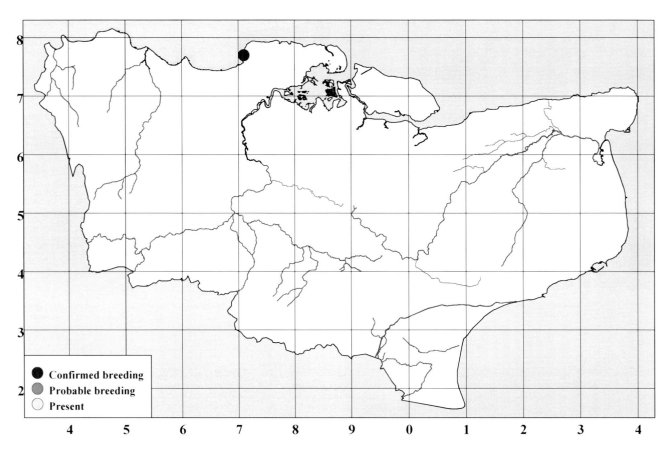

Confirmed breeding
● Probable breeding
○ Present

Description

Lestes viridis is very similar to our two resident lestid species. The male is a dark metallic green but does not develop the powdery blue pruinescence on the abdomen. The superior anal appendages are yellow with black tips and the inferior anal appendages are black and very short, being less than half the length of the superior anal appendages. The female is also dark metallic green. The pterostigmata are pale. Reference to a good field guide of European dragonflies is recommended, to be sure of a positive identification of any lestid species seen without any blue pruinescence.

Habitat

The preferred habitat for *L. viridis* is ponds, lakes, slow flowing rivers (Askew, 1988) and stagnant waters (d'Aguilar, Dommanget & Préchac, 1986). Overhanging trees and shrubs are required for oviposition, especially willows, but several other species have also been recorded including Hawthorn and some fruit trees.

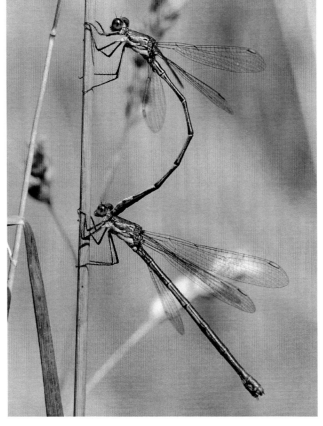

Willow emeralds, in tandem (Malcolm Farrow)

28

Breeding

The female inserts eggs into bark of overhanging twigs and branches of waterside trees causing some damage that may still be visible for two or three years after. The exuviae from Cliffe were found on a single occasion on a rush at the margin of a dyke. The exuviae have a wide labium unlike the 'racket' shaped labium of all other European members of the genus *Lestes*.

Distribution

L.viridis occurs in southern and central Europe as far north as Belgium and Germany and extending to the Middle East and North Africa. It is a very rare migrant to Britain with only one other record, that of a male adult captured at Shenley, Hertfordshire, on 11 August 1899 by E.R. Speyer (Merritt, Moore & Eversham, 1996). Some authorities have questioned the validity of this record as Speyer had also been collecting specimens in Europe during August 1899. There is also an unconfirmed and unpublished record of an adult caught near Hastings in Sussex during the 1980's (pers. comm.. D.Chelmick). In Kent, the only record is of an exuviae of *L. viridis* collected by the authors at Cliffe marshes on 29 June 1992, the identification of which was verified by British Dragonfly Society experts Graham Vick and David Chelmick.

Similar species

Lestes sponsa and *Lestes dryas* – the mature male *L.viridis* does not have any blue pruinescence.

Flight period in Kent

August and September; June to November in southern Europe (Askew, 1988).

Willow emerald, female (Hania Berdys, www.gardensafari.net)

Willow emerald, male (Malcolm Farrow)

Southern Emerald
Lestes barbarus (Fabricius)

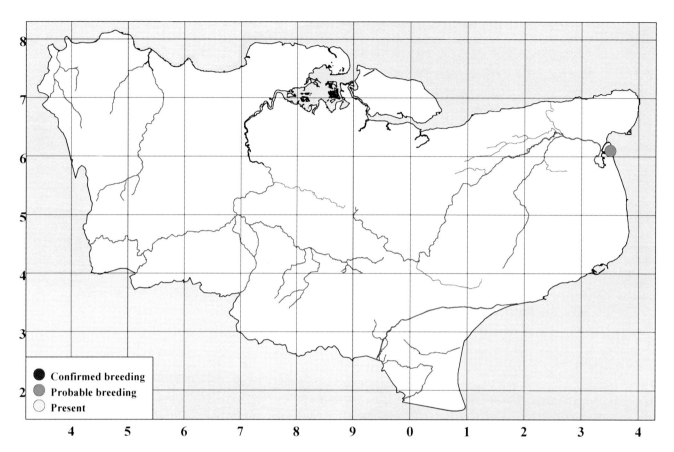

Description

The male *Lestes barbarus* is dark metallic green sometimes with a small amount of blue pruinescence on abdominal segment 10 and also on the thorax between the wing roots. The eyes are green. The inferior anal appendages are very short and each has a sharp point which is turned outwards. The female is dark metallic green with no blue pruinescence. Her eyes are green. In older specimens of both sexes the body colour may change from dark green to copper. The pterostigmata are *bi-coloured*, brownish on the inner half and creamy white on the outer half. In common with other Lestid species the wings are usually held half open when at rest.

Habitat

This species is found in stagnant or slightly brackish water (Askew, 1988) and in shallow and open ponds (Gerken & Sternberg, 1999). The known habitats in Britain are in coastal dunes. The present Kent location is a dune slack with shallow non-permanent water, which is densely vegetated with Sea Clubrush, *Bolboschoenus maritimus* (Linnaeus).

Southern Emerald, male

Breeding

The female lays her eggs, usually *in tandem* with the male, into stems of rushes such as Juncus and Carex spp.. She may also oviposit in branches of shrubs (Askew, 1988). Due to high habitat fidelity (Gerken & Sternberg, 1999) this species tends to remain faithful to the pond in which it developed (Utzeri *et al.*, 1984), therefore colonisation is slow. The eggs hatch the following spring and develop in two months (d'Aguilar *et al.*, 1986)

Distribution

In Europe this species has a mainly Mediterranean distribution, being widespread in Spain, Southern France and Italy. It becomes progressively scarce and local northwards. The most northerly record is from the southern tip of Sweden. In Britain it was first reported from Winterton Dunes, Norfolk in 2002. The following year, 2003, it was also seen and photographed near Sandwich in Kent. In 2004 up to 15 individuals were seen again near Sandwich and pairs were observed mating and ovipositing. Due to this site being flooded by sea water during an exceptional high tide, its status in Kent is now uncertain as no adults were found in 2005.

Similar species

Females of *L. sponsa* and *L. dryas*, and also immature males of these species before the blue pruinescence develops. However *L. barbarus* is distinguished by bi-coloured pterostigmata and green eyes.

Flight period in Kent

End of May to end of September in Europe.

Southern Emerald, female

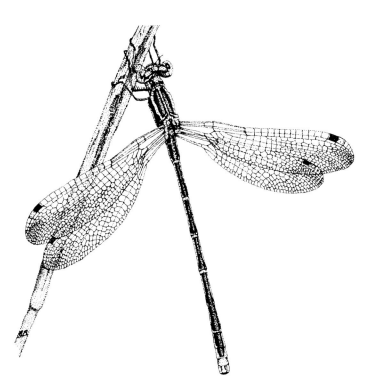

Emerald Damselfly

Lestes sponsa (Hansemann)

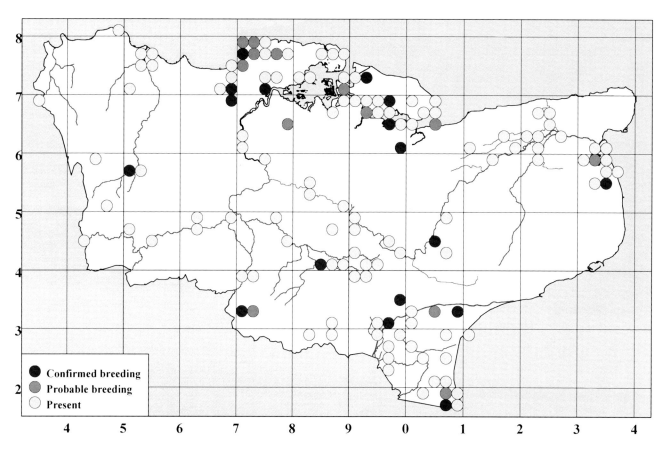

- ● Confirmed breeding
- ● Probable breeding
- ○ Present

Emerald Damselfly, male (Steve Smith)

Description

The mature male of this species is dark metallic emerald green with some bronze highlights on the abdomen. There is a powder blue pruinescence on segments 1 and 2, and also on 9 and 10. The eyes are blue. The female is more heavily built than the male and has no blue markings. Both sexes have clear wings with dark pterostigmata, which are narrower than those of *L. dryas*. This species is also less robust. Unlike most other damselflies, the species of this genus rest with wings held half open.

Habitat

This species can be found in a variety of waterbodies, from acid to brackish, such as ponds, ditches, lakes, pools and canals. Dense emergent and marginal vegetation is usually present.

Breeding

While the pair are still *in tandem*, the female oviposits into emergent plants such as rushes and reeds. She often works her way down the plant stem until she is completely submerged. If she climbs low enough even the attached male may become submerged. The eggs hatch the following spring and develop rapidly to emerge in 3 months or less. Exuviae may be found in large numbers in marginal plants.

Distribution

Lestes sponsa is found throughout Britain and is often abundant. In Kent its distribution seems to be divided into three distinct areas. One is in a fairly narrow northern band, from Cliffe through the Swale area and Stodmarsh to Sandwich, concentrated in marshland habitats; secondly, a more scattered Wealden distribution mostly in ponds and larger waterbodies, and thirdly, the dykes of the Romney and Walland Marshes.

Status

Regionally important in VC16

Similar species

Lestes dryas.

Flight period in Kent

Mid May to late September.

Emerald Damselfly, female (Nick Donnithorne)

Scarce Emerald Damselfly

Lestes dryas Kirby

Suborder Zygoptera

Family Lestidae

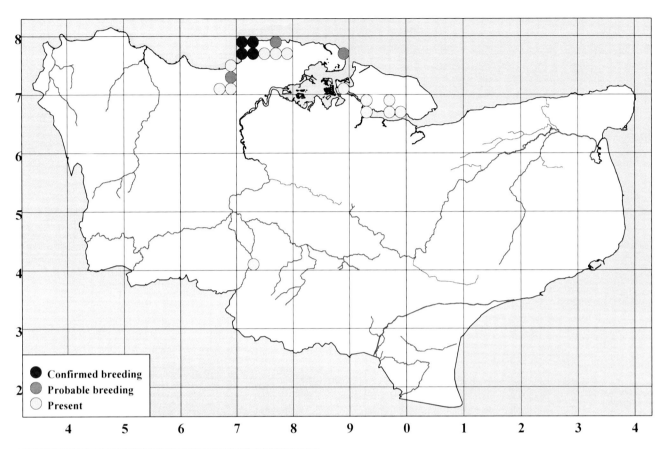

Confirmed breeding
Probable breeding
Present

Description

The mature male is basically a dark metallic emerald green with some blue markings. There is a powder-blue pruinescence on segment 1 and the first half of segment 2, although this can be variable, and also on segments 9 and 10. The eyes are bright blue. The female is also a similar metallic green with no blue markings. Segment 1 is pale with two square green spots. The female is more heavily built than the male. Both sexes have clear wings with dark, almost square pterostigmata. Both British *Lestes* species rest with wings held half open.

Habitat

This species is found by fresh to brackish shallow water in ditches on coastal grazing marshes. Emergent vegetation is dense and often the dominant species is Sea Clubrush *Bolboschoenus maritimus* (Linnaeus). It is also occasionally found inland at small densely vegetated shallow ponds.

Scarce Emerald Damselfly, male

34

Breeding

The pair are usually *in tandem* while the female oviposits into plant stems. The eggs overwinter and hatch very early in spring, then develop quickly to emerge in late June on marginal plants. Exuviae have been found in dykes on Cliffe Marshes and probable breeding was recorded from three other tetrads nearby where larvae were found.

Distribution

Lestes dryas is a rare species that was thought to be extinct in Britain in 1971 (Hammond, 1983). It is now found around the Thames Estuary and inland in Norfolk. In Kent there are early records from V.C. 15 (Longfield, 1949a), and Norman Moore recorded this species from a ditch in Kent near Camber on 17th July 1945 and also from a ditch near Wittersham on 9th June 1946. A mounted *L. dryas* from Maidstone Museum was labelled "Romney Marsh August 1st 1952". Unfortunately this specimen was damaged in a fire at the museum and its label did not have the recorders name on it. This species was rediscovered in 1983 in Kent at Thong pond (Kent Biological Archives and Records Centre). In 1991, a number of larvae were discovered in dykes at Cliffe marshes by Eric Philp, and it has since been reported eastwards to the Hoo Peninsular and also from Elmley Island on the Isle of Sheppey; in 2004, the species was found at Shorne Wood Country Park. The most recent new location was near Horsmonden, in the south of the county, where a female was photographed in 2008.

Scarce Emerald Damselfly, female

Status

Nationally important

Similar species

Lestes sponsa.

Flight period in Kent

Late May to mid August.

White-legged Damselfly
Platycnemis pennipes (Pallas)

Suborder Zygoptera
Family Platycnemididae

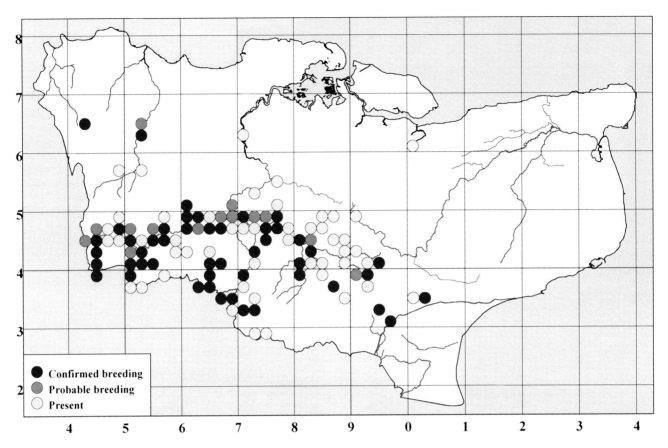

Confirmed breeding
Probable breeding
Present

Description

This damselfly is the only representative of its genus in Britain. The male has a pale creamy-green and black thorax giving it a striped appearance. The abdomen is pale blue with a black dorsal line on segments 1-5, a pair of spots on 6, and paired linear black markings on 7-10. The immature of both sexes have a pale creamy-white abdomen. The female at this stage is known as 'form *lactea*' and her abdomen has paired spots on segments 2-6 and paired linear markings on 7-9. As she matures the paired spots develop into paired linear markings and her colour becomes pale creamy-green. The wings of both sexes are clear with short brownish pterostigmata. The most characteristic feature of this species is the legs, which have broad white tibiae with long hairs, suggesting a feather-like appearance.

Habitat

It was thought that this species preferred slow flowing rivers with dense marginal vegetation. It has recently also been found breeding in lakes and large ponds, most of which were still waters with no inflow.

Very often it can be found, in abundance, in adjacent areas of long dense grasses and herbage.

Breeding

With the male still attached, the female, often accompanied by numerous other pairs, oviposits into submerged plants with her abdomen often partly submerged. The male remains motionless and vertical with wings closed in the 'sentinel' position. When ready to emerge the larvae climb onto marginal vegetation, sometimes a metre or two away from the water. The exuviae have a distinct thread-like point on the lamellae.

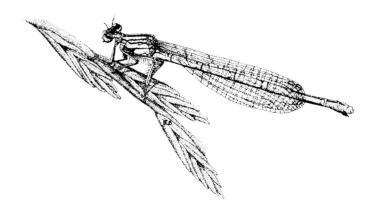

Distribution

This damselfly has a scattered distribution in southern England south of The Wash, mostly near lowland rivers and canals. It can be locally abundant and often out-numbers other species present. In Kent, it has a south-western distribution especially along the Rivers Medway and Beult, and is very common within this area. A very surprising isolated breeding was recorded by the authors on the River Darent near Lullingstone Castle. Cynthia Longfield could not understand why it had not been found on this river (Longfield, 1949b). *Platycnemis pennipes* was reported to be "entirely a running water species in the British Isles" (Longfield, 1949a). Although this species prefers slow flowing rivers, recent records in Kent show that this species also breeds in many large ponds and lakes with no inflow or outflow. The authors have collected many exuviae from farm ponds in the Weald, one of which had little marginal or emergent vegetation.

Status

Regionally important in VC15

Similar species

Enallagma cyathigerum - pale immatures resemble the adult *P. pennipes.*

Flight period in Kent

Late April to late September.

Male and female White-legged Damselfly, in the wheel position

Large Red Damselfly

Pyrrhosoma nymphula (Sulzer)

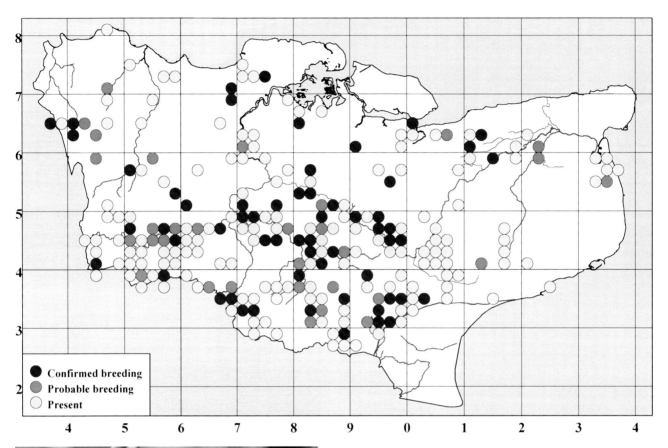

● Confirmed breeding
● Probable breeding
○ Present

Description

This damselfly is easily identified by its red and black colour. The male has a black thorax with some yellow on the lower sides, and black legs. The two broad antehumeral stripes, the abdomen and the eyes, are red. Segments 7- 10 have varying amounts of black on them. The clear wings have black pterostigmata. Females are usually very similar, the main difference being in the variable amount of black on the abdomen and a narrow band of yellow at each segment joint. One form f. *intermedia*, has a black dorsal stripe, black bands adjacent to the yellow bands, and mostly black segments 6-10. Another form, f. *fulvipes* Stephens, has a finer black dorsal stripe and reduced amounts of black on the segments. A third form, f. *melanotum* S élys, has yellow antehumeral stripes and a mostly black abdomen with yellow segment joints. This species is usually the first to be seen in spring.

Large Red Damselfly, male and female in tandem

Habitat

This damselfly can be found in a wide variety of habitats including ponds, lakes, ditches, rivers and canals. It also occurs in acid bog pools and brackish water and may often be found quite far from water in woodlands, hedgerows and grasslands.

Breeding

Eggs are laid into the tissue of submerged or floating plants usually while the pair are *in tandem*. Larval development may take two years or more and when completed, the larvae climb onto marginal plants often rushes (*Juncus* spp.). The exuviae are usually dark brown with an 'X' mark on the lamellae, if visible, and the rear of the head has a square outline.

Large Red Damselfly, male (Steve Smith)

Distribution

Pyrrhosoma nymphula is very widespread and common over almost the whole of Britain. In Kent, it has mainly a southern distribution with fewer sites in the northern half of the county. This may seem very surprising in view of its widespread and common national status. Although it is showing signs of decline in intensively cultivated areas, it may be that it has been under-recorded. One early record for this species comes from Henry Ullyett who, in 1877 saw many of these "red damsels" at a large two or three year old pond which was formed by the collapse of a tunnel at Folkestone Warren (Ullyett, 1888).

Similar species

Small Red Damselfly, *Ceriagrion tenellum* (de Villers) - which is much smaller, has red legs and is not found in Kent.

Large Red Damselfly, female

Flight period in Kent

Late June to mid September.

Red-eyed Damselfly

Erythromma najas (Hansemann)

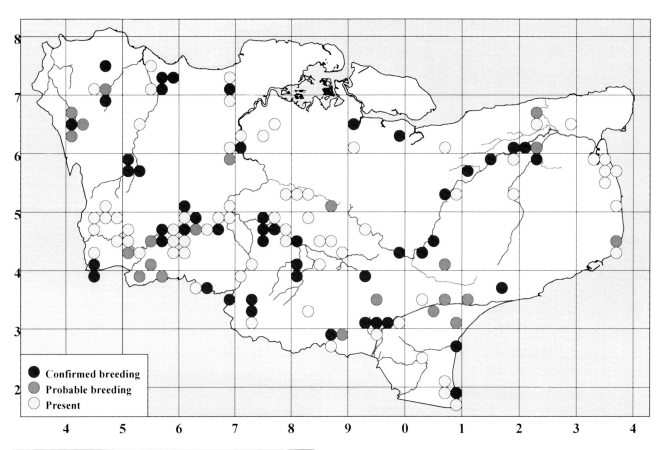

Description

The male of this species has a black thorax with sky-blue sides, and a grey-black abdomen. Segments 1 and 9-10 are sky-blue. The blood red eyes are conspicuous. Females are longer bodied and more robust with brownish eyes. The black thorax has greenish sides and reduced yellow antehumeral stripes, and the abdomen is black. The clear wings of both sexes have pale squarish pterostigmata. These damselflies spend much of their time over the water where there are plants with floating leaves and so are often observed from a distance.

Habitat

Still waterbodies, such as lakes and ponds, and slow flowing rivers, streams and canals, with floating vegetation, are typical sites for this species. Large leaved aquatic plants, such as all types of Water Lilies, and Broad-leaved Pondweed *Potamogeton natans* (Linnaeus) are especially preferred.

Red-eyed Damselfly, male (Lee Manning)

40

Breeding

The pair usually remain *in tandem* while the female oviposits into the underside of floating leaves. Often they become completely submerged if plant stems are chosen where floating leaves are not present. The robust larvae take up to two years to develop. Exuviae may be found on rushes and reeds at the margin and are readily identifiable by their caudal lamellae which have rounded tips and three dark transverse bands beyond the mid-way node.

Distribution

Erythromma najas has a mainly southern distribution from The Wash to the Welsh borders. It is also found, but less frequently, in the West Midlands and Yorkshire areas. In Kent it has a widely scattered distribution. There are pre-1910 records from V.C.16 (Longfield, 1949a) and this species was certainly present at Keston in 1949 (Longfield, 1949b).

Similar species

Ischnura elegans - which is a smaller and more delicate species that does not have red eyes.

Erythromma viridulum - which is a smaller species with a black 'X' mark on segment 10 and blue on the sides of segment 8.

Flight period in Kent

Early May to mid September.

Red-eyed Damselfly, female (Lee Manning)

Small Red-eyed Damselfly

Erythromma viridulum (Charpentier)

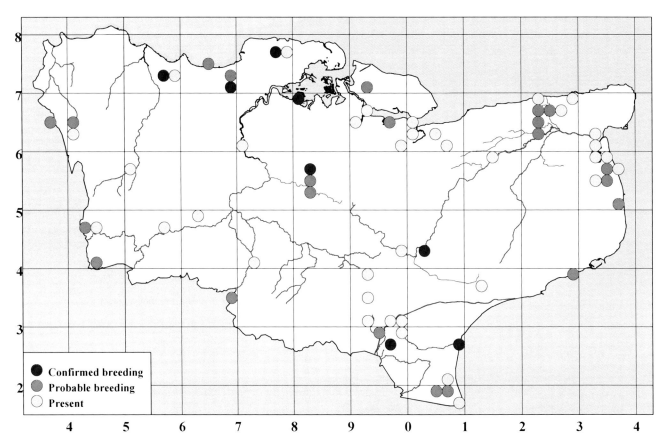

Legend:
- ● Confirmed breeding
- ● Probable breeding
- ○ Present

Description

The male has a black thorax with blue sides and with short fine yellow antehumeral stripes that are only visible at very close range. The abdomen is metallic black except for segments 1 and 9-10 which are bright blue, but S10 also has some black on it which is often described as an 'X' mark. This mark is variable and not obvious until seen at close range. When viewed from the side S8 has blue on the lower half, but is black dorsally. The eyes are tomato red. The female has brownish eyes and has no blue abdominal segments. Both sexes spend much time flying over or resting on vegetation far from the margin and so are usually observed from a distance.

Habitat

E.viridulum occurs on still waterbodies such as dykes, ponds and lakes and will tolerate slightly brackish waters. These damselflies seem to show a preference for areas with mat-forming plants such as Canadian Pondweed *Elodea* spp., Water-milfoil *Myriophyllum* spp., Hornwort *Ceratophyllum* spp. and surface algae. It has been reported where the vegetation included the pondweed *Potomageton natans* L., Amphibious Bistort *Polygonum amphibium* L., water lilies and other floating-leaved plants.

Breeding

The pair remain in tandem while the female oviposits into plant tissue just below the surface. Emergence takes place after one year with the larvae climbing onto rushes and sometimes dead branches of trees in the water near the margin. The exuviae are small and lack the bands of pigmentation that are seen on the lamellae of *E. najas.* Examination under a magnification of x 20 is necessary in order to separate the exuviae of this species from other coenagrionids such as *Enallagma cyathigerum.*

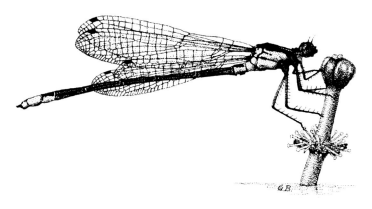

Distribution

This species was first discovered in Britain in July 1999 in the Bradwell-on-Sea area of Essex. It was reported from the Isle of Wight in 2000 and immigration continued into other counties of the south-east in 2001 including Norfolk, Bedfordshire and Kent. The colonisation of *E. viridulum* is the first of its kind. No other migrant damselfly species has established widespread colonies in the British Isles (Cham, 2002). In Kent it was observed at three localities in 2001, one of them being the perimeter lake at Bluewater Shopping Centre, Greenhithe, near Dartford, where the first exuviae of this species were collected by the authors in July 2002. From the locations near the coast it moved inland to places such as Scotney Castle, Lamberhurst and Haysden Country Park, Tonbridge, and is now widespread throughout Kent.

Similar species

E. najas – which is larger, the male of which has no blue on S8.

Ischnura elegans – which is similar in size but does not have red eyes and has two bright blue post-ocular spots which are absent in *E. viridulum*. Only S8 is blue on *I. elegans* and the wings have bi-coloured pterostigmata.

Flight period in Kent

Late June to mid September.

Small Red-eyed Damseflies, ovipositing

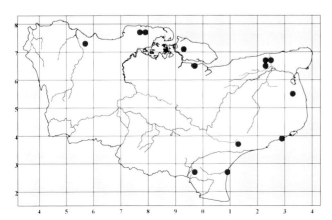

Records of Small Red-eyed Damselfly to 2002

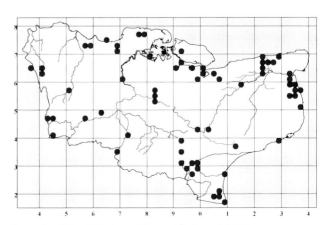

Records of Small Red-eyed Damselfly to end 2008

Azure Damselfly
Coenagrion puella (Linnaeus)

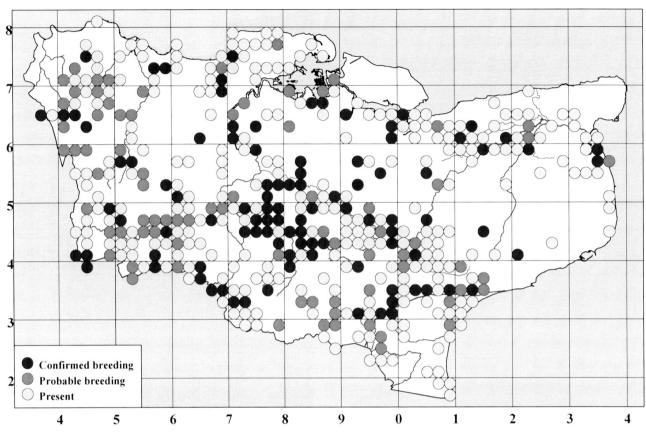

- ● Confirmed breeding
- ● Probable breeding
- ○ Present

Azure Damselfly, male

Description

The male of this species has a blue thorax with a black dorsal surface and blue antehumeral stripes. The abdomen is also blue with black markings on segments 1-7 and 9 and 10. Segment 8 is all blue. The shape of the marking on segment 2 resembles a glass tumbler or 'U' but can sometimes be variable. Most females have a green abdomen with black markings on all segments which almost obscure the green and so appear almost completely black. A few females have a blue abdomen with a black dorsal surface which is broken by 3 or 4 blue bands. Both female forms may have an elongated thistle-head marking on segment 2. The clear wings of both sexes have dark pterostigmata.

Habitat

Coenagrion puella may be found at many types of waterbodies including ponds, lakes, canals, dykes and even small garden ponds, with plenty of marginal and emergent plants.

Breeding

Oviposition usually takes place with the male and female *in tandem*. While she oviposits into aquatic vegetation, the male will hold himself vertically in the sentinel position with wings closed. When ready to emerge the larvae will climb a suitable plant support at or near the water margin. Exuviae are quite easy to identify as they are strongly peppered with dark spots behind the eyes, and the lamellae are usually bluntly pointed.

Distribution

Coenagrion puella is common and widespread over most of Britain but is absent in the north of Scotland. In Kent it is very common and widespread. On smaller waterbodies it is more common than *Enallagma cyathigerum*.

Similar species

Enallagma cyathigerum - the male has a mushroom shaped mark on segment 2 and segment 9 is all blue.

Coenagrion pulchellum - the male has a wine glass shaped mark on segment 2.

Flight period in Kent

Late April to mid September.

Azure Damselflies in the mating wheel

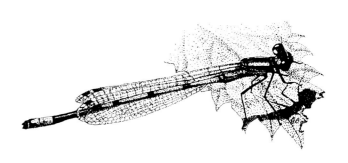

Variable Damselfly

Coenagrion pulchellum (Vander Linden)

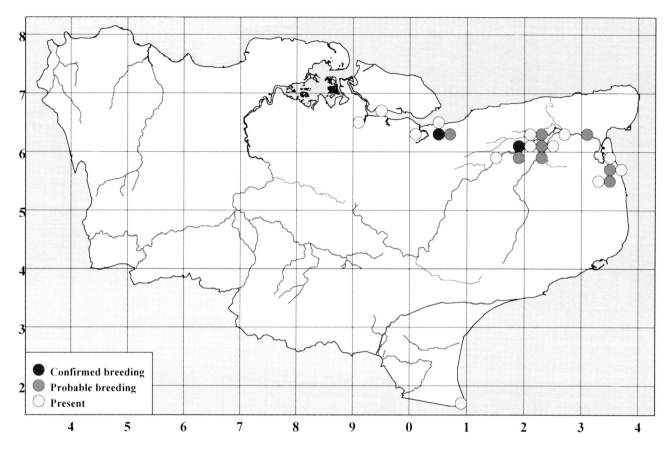

Confirmed breeding
Probable breeding
Present

Description

The male *C. pulchellum* has a blue thorax with a black dorsal surface and broken blue antehumeral stripes resembling exclamation marks. The abdomen is blue with black markings on segments 1-7 and 9 and 10. Segment 8 is all blue. The shape of the marking on segment 2 is like a wine glass and is variable. Sometimes the stem of the glass is absent or almost so, making the marking as that of *C. puella*. The female of this species has a blue abdomen with more extensive black markings than the male showing only a few bands of blue on the dorsal surface. A darker female form appears completely black with the exception of a characteristic marking on segment 2 that is common to both female forms. Both sexes have clear wings with pale pterostigmata.

Habitat

Still or slow flowing waterbodies best suit *C. pulchellum*. Ponds and dykes with emergent and marginal vegetation are preferred.

Breeding

The pair usually remain *in tandem* while the female oviposits into submerged vegetation. She may become submerged while doing so. The larvae probably complete their development in one year but may take longer in more northerly areas of Britain. Exuviae may be found among the marginal vegetation and are very similar to those of *C. puella*. The lamellae usually have rounded ends with no point.

Variable Damselfly, male

Distribution

Coenagrion pulchellum has a very scattered and fragmented distribution in Britain. It occurs mainly in southern England, East Anglia, South Wales and southwest Scotland. In Kent, it appears to be limited to the marshes and levels in the north-east of the county at such places as Graveney, Westbere, Hacklinge and Preston Marshes where aquatic and marginal vegetation is dense. It may be more widespread in the levels and marshes of the Rivers Wantsum, Great and Little Stour than records show. There are early records for *C. pulchellum* at Deal by C. G. Hall, and at Canterbury by H. M. Briggs (Lucas, 1900). It was also recorded at Ramsgate (Page, 1908), and at Newenden by Norman Moore on 9th July 1946. However Longfield thought that it had died out in Kent (Longfield, 1949a).

Status

Nationally important

Similar species

Coenagrion puella - the male has a tumbler shaped mark on segment 2 and unbroken antehumeral stripes.

Enallagma cyathigerum - the male has a mushroom shaped mark on segment 2, and segment 9 is all blue.

Flight period in Kent

Early May to mid August.

Variable Damselfly, female (Nick Donnithorne)

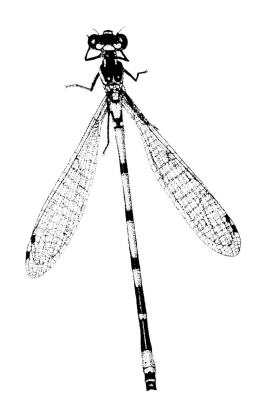

Common Blue Damselfly

Enallagma cyathigerum (Charpentier)

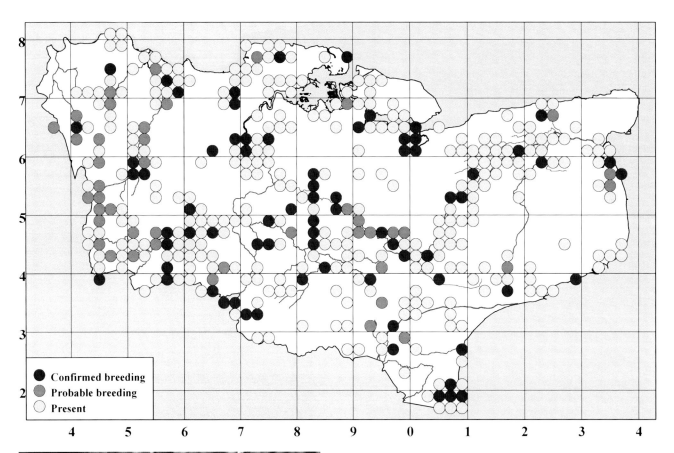

Legend:
- ● Confirmed breeding
- ● Probable breeding
- ○ Present

Common Blue Damselfly, male

Description

The male of this species has a blue thorax with a black dorsal surface and blue antehumeral stripes. Its abdomen is also blue with black markings on segments 1-7 and 10, but segments 8 and 9 are completely blue. The shape of the marking on segment 2 resembles a mushroom, although this can be variable. Females have elongated black dorsal markings on all segments on a blue, yellow-brown or greenish abdomen. All female colour forms have a thistle head marking on segment 2. The clear wings of both sexes have dark pterostigmata.

Habitat

Enallagma cyathigerum can be found at a great variety of still or slow flowing waterbodies such as ponds, lakes, and canals. It can also be found in acid waters.

Breeding

Oviposition may take place while the pair are *in tandem* or the female may be alone. She oviposits into a variety of submerged plants and may become submerged while doing so. If the male has been *in tandem*, he may release the female at this stage. The larvae usually take one year to develop, but often longer in more northerly areas of Britain. When development is complete the larvae usually climb plants anywhere in the water or at the margin. The authors found numerous exuviae at one site in Scotland where the larvae had crawled about six metres from the water and had then climbed about two metres up a bank to emerge on grasses.

Distribution

As its name implies, this species is common and widespread over the whole of Britain. In Kent it is also common and widespread but in the smaller habitats such as garden ponds, it is often absent or less common than *Coenagrion puella*, whilst at lakes it may be seen in hundreds, as at Milton Lakes near Canterbury.

Similar species

Coenagrion puella - the male has a tumbler shaped marking on segment 2.

Coenagrion pulchellum - the male has a wineglass shaped marking on segment 2.

Flight period in Kent

Late April to early October.

Common Blue Damselfly in the mating wheel

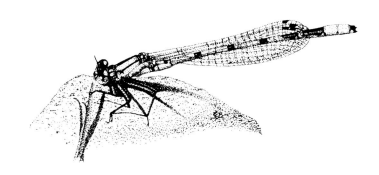

Blue-tailed Damselfly

Ischnura elegans (Vander Linden)

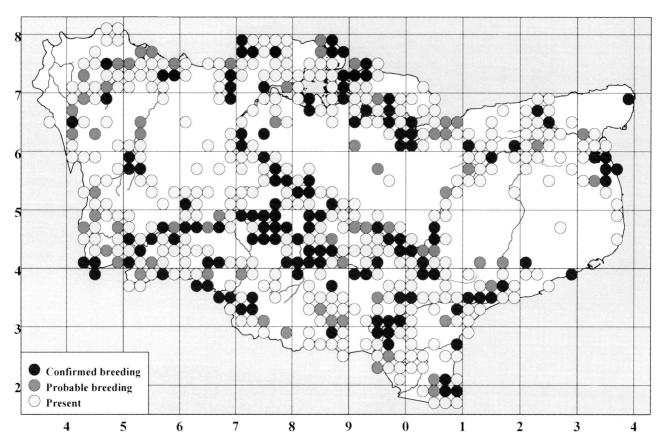

- ● Confirmed breeding
- ● Probable breeding
- ○ Present

Description

Males of this species have a blue or green thorax with a black dorsal surface and blue or green antehumeral stripes. The abdomen is black with a bright blue segment 8. Females that are similar to the males are known as f. *typica* (Vander Linden) and the immature form of this is f. *violacea* Sélys which has a violet thorax and a blue segment 8. However, some f. *violacea* individuals mature to become f. *infuscans* Campion which has a greenish thorax and brown segment 8. The immature f. *rufescens* Stephens has a pink thorax with only a black mid dorsal stripe and a blue segment 8. This form matures to become f. *rufescens-obsoleta* Killington which has a brownish thorax and a brown segment 8. All these forms occur in Kent. The pterostigmata are two-coloured, black and cream.

Habitat

This very common species may be found at ponds, ditches, lakes, canals and slow flowing streams and rivers. It will tolerate some pollution and brackish water too. New ponds will quickly be colonised by this pioneering species which often travels long distances from water.

Blue-tailed Damselfly, male (Lee Manning)

50

Breeding

The females of *I. elegans* oviposit on their own into the tissue of various aquatic plants. In the south, larval development takes only one year but may take longer in northern Britain. The exuviae can be found on emergent and marginal vegetation at or near the waters edge. The lamellae of the exuviae are long and pointed.

Distribution

This is a very common and widespread species over almost the whole of Britain. In Kent it is also very common and widespread, being found in about 65% of the tetrads where dragonflies were recorded.

Similar species

Small Blue-tailed Damselfly, *Ischnura pumilio* (Charpentier 1823) - which is smaller with part of segment 8 and all of segment 9 being blue. It is not found in Kent.

Erythromma najas - which is more robust and has blood red eyes.

Erythromma viridulum - which has red eyes.

Flight period in Kent

Mid April to early October.

Blue-tailed Damselfly, female, form *violacea* (Lee Manning)

Blue-tailed Damselfly, female, form *rufescens* (Steve Smith)

Migrant Hawker

Aeshna mixta Latreille

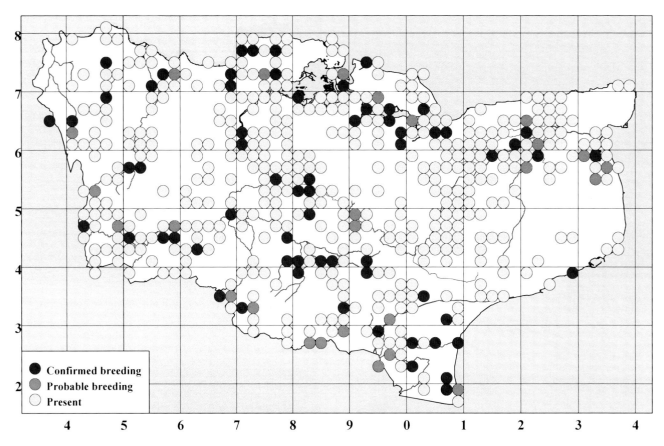

Migrant Hawker, male

Description

A medium sized hawker dragonfly with a brown thorax and brown to black-brown abdomen. The thorax has lateral stripes that blend from yellow to blue, and pale yellow antehumeral stripes that are almost reduced to spots. Both sexes have an elongated cream marking on segment 2, which in the male reaches a blue band. The male has two blue spots on segments 3-10, and has blue eyes. The female's spots are yellow and the eyes are brownish. Both sexes have clear wings with long brown pterostigmata. They are more tolerant of their own kind than other hawkers and may often be seen together in large numbers as they hunt for food.

Habitat

This species is found in a variety of static and slow-moving waterbodies including brackish, but not acidic, water with plenty of marginal vegetation. Also frequently found at woodland glades, woodland edges and hedgerows.

Breeding

The female deposits her eggs in plant stems or mud at the waters edge. After over-wintering the larvae hatch in early spring and development is quite rapid, allowing emergence in the same year. The larvae will then climb a suitable support on which to emerge close to the waters edge. Exuviae are found on rushes, reeds or other plant stems, often in good numbers.

Distribution

The Victoria History of the County of Kent (Page, 1908) states that "of the rare *Aeshna mixta*, Kent has produced a good number during the last year or two" at many east coastal locations. In the 1940s, *A. mixta* was an uncommon migrant from southern Europe (Brooks, 1997), and was reported as a regular immigrant in the south and southeast (Longfield, 1949a) and well established and breeding in some localities (Longfield, 1949b). It is now a resident species, becoming increasingly common over most of England, from the Humber southwards, and in southern Wales. In Kent it is very common and widespread.

Similar species

Aeshna juncea - not found in Kent.

Aeshna cyanea - which has broad green antehumeral stripes and green abdominal markings.

Flight period in Kent

Early July to mid November.

Migrant Hawker, female (Bryan Bullen)

Southern Hawker
Aeshna cyanea (Müller)

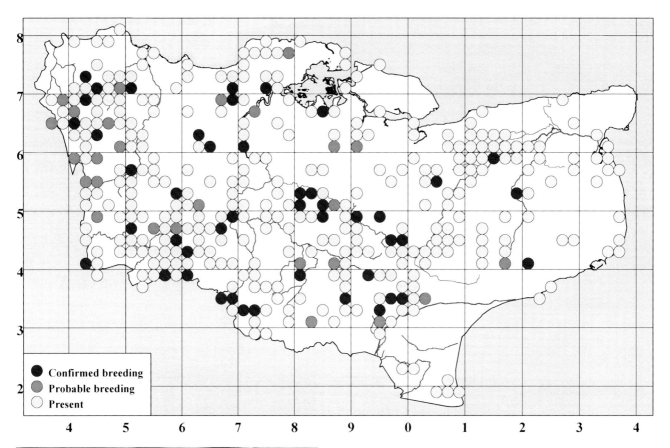

Southern Hawker, female

Description

A large dragonfly with a dark brown thorax and abdomen. The thorax has broad apple green antehumeral stripes, and the abdomen of the male has green spots on segments 1-7. Segment 8 has blue spots and 9-10 have blue bands. Segment 2 has an elongated yellow marking on the mid-dorsal line. The female is similar with the exception of green markings on 8-10, not blue. The clear wings develop an amber tint with age and the pterostigmata are dark brown or black.

This is an inquisitive and territorial species which will often 'inspect' the observer at close quarters.

Habitat

Aeshna cyanea may be frequently seen patrolling woodland ponds or lakes and at garden ponds. It also frequents woodland rides, clearings and hedgerows in search of prey.

Breeding

The female deposits her eggs, often accompanied by the sound of rustling wings, in rotten wood, soft mud, moss or living or dead plant material, which may be in the water or close by, or sometimes a distance away. When development is complete the larva will climb a suitable support from which to emerge. Exuviae are typically found on reeds or rushes right at the waters edge, and may also occasionally be found hanging from low branches close to the water surface.

Southern Hawker, female, egg-laying

Distribution

Aeshna cyanea is common in lowland areas of southern England but scarcer in the north. It is also widespread in Wales and a few isolated populations occur in Scotland. In Kent, it is common and widespread, however, perhaps due to its territorial behaviour, it is usually only seen in low numbers. For example, although only one adult was seen at a densely vegetated pond near Tunbridge Wells during a visit one August, the exuviae of this species were so numerous and easily visible that thirty were gathered in a few minutes.

Similar species

Aeshna juncea - not found in Kent.

Aeshna mixta - has blue or yellow abdominal markings and much reduced antehumeral stripes.

Brachytron pratense - is smaller and has blue or yellow abdominal markings. It has an earlier flight period.

Flight period in Kent

Late May to mid November.

Brown Hawker

Aeshna grandis (Linnaeus)

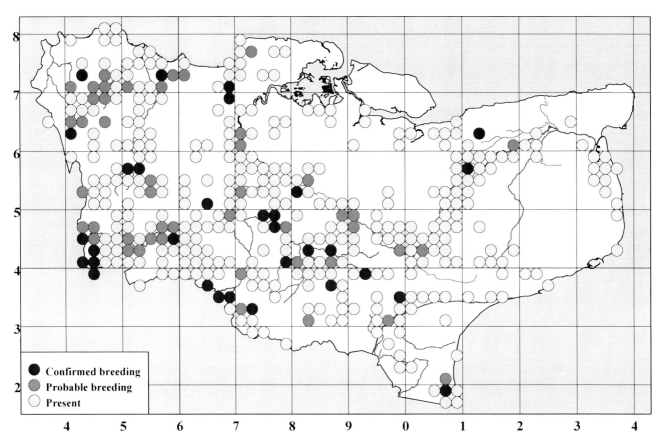

Confirmed breeding
Probable breeding
Present

Description

A large dragonfly with brown thorax and abdomen. The thorax has two yellow lateral stripes but no antehumeral stripes. The male has two blue spots on segment 2, blue markings on the sides of the abdomen and bluish eyes. The female has pale yellow abdominal markings and yellowish eyes. Both sexes have amber wings with long brown pterostigmata. This easily identifiable dragonfly is well camouflaged and difficult to spot when at rest.

Habitat

Aeshna grandis may be found at almost any static or slow flowing waterbody from garden ponds to lakes, from canals to slow rivers, from ditches to gravel pits, and where water is quite polluted.

Breeding

The female will lay her comparatively large eggs, often in rotten wood, in any of the above types of waterbodies. The eggs overwinter and the larvae hatch in spring. Larval development may take two to four years and on completion the larvae will climb up and emerge on rushes or reeds out over the water or at the waters edge. It is here where the exuviae may be found.

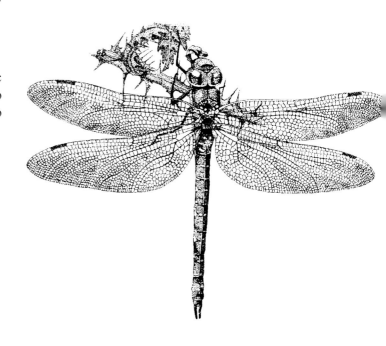

Brown Hawker, female, egg-laying

Distribution

Aeshna grandis is common and widespread in lowland areas of England and parts of North Wales. In the past it was considered very common in Kent (Longfield, 1949a), and it is still true of this species today.

Similar species

Norfolk Hawker, *Aeshna isosceles* (Müller) - which has clear wings and green eyes, but is not found in Kent.

Flight period in Kent

Late May to early October.

Emperor Dragonfly
Anax imperator Leach

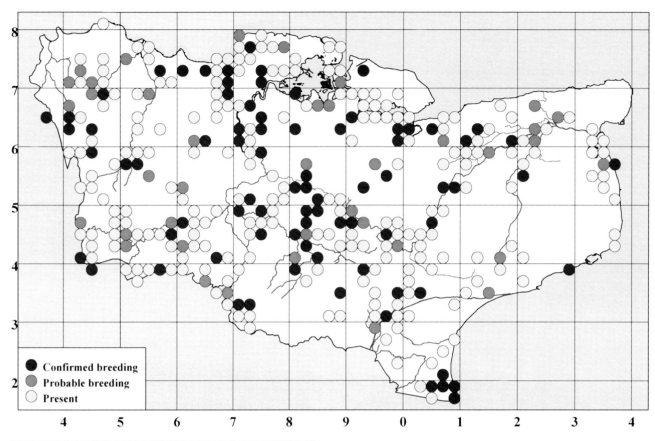

● Confirmed breeding
● Probable breeding
○ Present

Description

A large dragonfly with an apple green thorax and no lateral or antehumeral stripes. The male has a deep blue abdomen with a broad black dorsal stripe and has blue-green eyes. The female's abdomen is green, sometimes with a hint of blue, with a broad black dorsal stripe, and she has greenish eyes. The clear wings have long narrow mid-brown pterostigmata. This is a very aggressive species with exceptional abilities in flight.

Habitat

Anax imperator is present at most static waterbodies from garden ponds to large lakes which are well vegetated with aquatic plants.

Breeding

The female oviposits in aquatic plants near the water surface and is often seen settled on a floating leaf with abdomen arched downwards into the water depositing her eggs into the leaf stem. Larval development usually takes one to two years and when complete the larvae climb a suitable support from

Emperor Dragonfly, pre-flight male

58

which to emerge. Emergence usually takes place after sunset and the maiden flight often takes place as dawn breaks. This species has also been seen emerging in the early daylight hours. Exuviae are typically found on stout reeds or rushes in the water and occasionally suspended from low branches overhanging the water surface.

Distribution

Anax imperator is widespread in low lying areas of southern England and southern Wales. In Kent it is common and widespread, though usually only seen in low numbers. This comment is applied to several of the aeshnid dragonflies. However, the following breeding records illustrate how prolific these species can be. At the author's garden pond (TQ85B) Leeds near Maidstone, in May 1992, a total of 94 exuviae of *A. imperator* were collected over a three day period, with a grand total of 144 for that season.

Flight period in Kent

Late April to late September.

Lesser Emperor

Lesser Emperor

Anax parthenope Sélys-Longchamps

Suborder Anisoptera, Family Aeshnidae

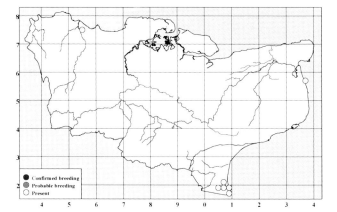

Description

This is a medium sized hawker dragonfly, smaller than *Anax imperator*. The thorax is green-brown and the abdomen is of similar colour, but darker, with a broad black stripe from segments 3-10. Segment 2 has a narrow yellow band at its junction with segment 1, and the rest of segment 2 and half of 3 is bright blue. The wings are lightly suffused with amber and the pterostigmata are long and yellow-brown. The eyes are green.

Status

This species, which breeds in small lakes and ponds, is a regular migrant to Britain from southern Europe. The first British record is from Gloucestershire in June 1996 (Brooks, 1997). In Kent individual males have regularly been recorded from the Dungeness area. The first was near the Bird Observatory on August 4th 1998 and was verified by a photograph, the second was from north of Boulderwall Farm on Denge Marsh, seen between 8th and 17th August 1998 and may have been the same individual (Parr, 1999). The third was from the Long Pits, Dungeness on 10th July 1999 (Attridge, 2000). Since then this species has been seen annually at Dungeness. All records were verified by the Migrant Dragonfly Project Odonata Records Committee.

Similar species

Anax imperator.

Hemianax ephippiger - a very rare migrant.

Hairy Dragonfly

Brachytron pratense (Müller)

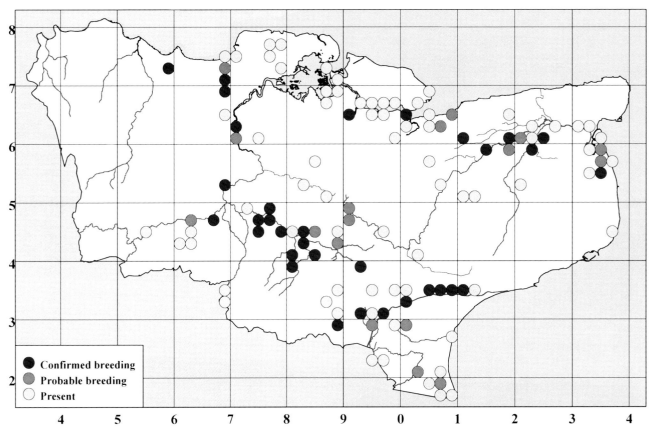

Confirmed breeding
Probable breeding
Present

Description

A medium sized hawker dragonfly with a dark brown thorax and black abdomen. The sides of the thorax of the male are apple green with similar coloured antehumeral stripes and the abdomen has paired blue markings on segments 2-10. The eyes are blue. The female has much reduced antehumerals which are yellow and the abdominal markings are also yellow. The female's eyes are brownish. Both sexes are noticeably downy, especially on the thorax. The pterostigmata are long and thin and of greyish-brown coloration. The wings are clear, becoming slightly yellowish in mature females. This is the first hawker species to emerge.

Habitat

Brachytron pratense favours still linear waterbodies, such as ditches, dykes and canals, but can also be found on ponds and lakes. They require abundant emergent and marginal vegetation with occasional trees.

Hairy Dragonflies, in the mating wheel (Mike Thurner)

Breeding

The female oviposits in plant material, favouring dead floating organic debris near the water's edge. The eggs hatch within a few weeks and when development is complete and the adults have emerged, the exuviae can be found close to, or at the water's edge, in dense rushes and reeds and can be difficult to spot.

Distribution

Brachytron pratense has a patchy distribution mainly in southern Britain with a few sites in western Scotland. In Kent, it also has a patchy distribution principally in the areas of the north Kent marshes, the Rivers Great Stour, Medway and Beult, the Royal Military Canal and Romney Marsh. Records appear to indicate that it is on the increase, though it may have been previously under-recorded due to its early emergence.

Hairy Dragonfly, male (Mike Thurner)

Status

Regionally important in VC16

Similar species

Aeshna cyanea - is larger and has green abdominal markings.

A. juncea -not found in Kent.

A. mixta - has a prominent yellow mark on abdominal segment 2, and blue/yellow on the side of the thorax.

Flight period in Kent

Mid April to mid July.

Golden-ringed Dragonfly
Cordulegaster boltonii (Donovan)

Suborder Anisoptera

Family Cordulegastridae

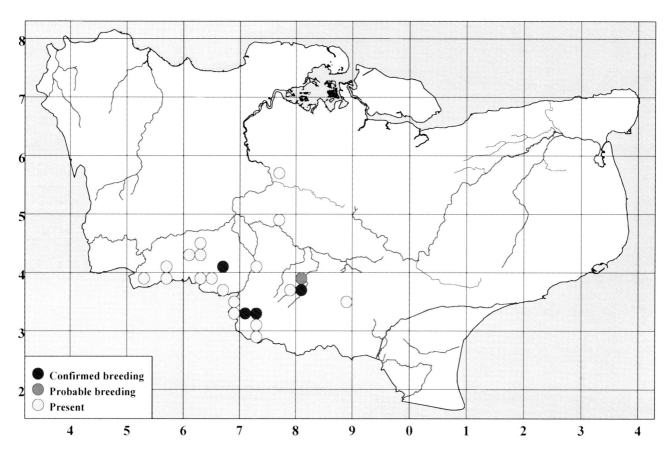

- ● Confirmed breeding
- ● Probable breeding
- ○ Present

Description

A striking and large dragonfly with black thorax and abdomen. The thorax has golden yellow lateral and antehumeral stripes and the abdomen has golden yellow rings. The male's abdomen is club shaped and the female has a long stout ovipositor extending beyond the abdomen making her the longest British species. The clear wings have black venation with the exception of the leading edge costal vein which is yellow. The pterostigmata are black, long and narrow, and the eyes are green. This species often rests hanging vertically from plant stems or branches.

Habitat

Cordulegaster boltonii, one of the largest dragonflies in Britain, can be found in the smallest of habitats. It can be seen patrolling small, often deeply cut, clear streams of moderate flow, often less than one metre wide. There may or may not be submerged aquatic vegetation and the margins may have shrubs and rushes, but usually not shaded by trees. A clear flight path is preferred along the watercourse in an open aspect such as heathland and suitable streams in woodland clearings.

Golden-ringed Dragonfly, female

Breeding

Whilst hovering alone, the female oviposits by stabbing her ovipositor repeatedly into sand or mud where the water is shallow. In older specimens, the ovipositor may be damaged and the lower half of the abdomen discoloured by mud or humic material (Brooks, 1997). After larval development, which may take two or more years, the adults emerge on convenient supports at the waters edge. Exuviae are typically found on, or suspended under, overhanging vegetation. One breeding site in Bedgebury Forest was confirmed by the authors when an exuviae was found in June 1994 and a larva seen in February 1996. Exuviae have now been found at Sissinghurst, Bedgebury Pinetum and at a site near Brenchley. These are woodland sites with suitable streams with an open aspect.

Distribution

Cordulegaster boltonii is common in southern, western and northern England and most of Scotland and Wales. In Kent it is confined, in small numbers, to the south-west of the county. Reports of this species have increased in frequency with adults having been recorded at several new locations. In a footnote in the Victoria History of the County of Kent, under the species name of *C. annulatus*, this species was thought to be "almost certainly present" (Page, 1908).

Status

Regionally important in VC15 and VC16.

Flight period in Kent

Late May to late August.

Downy Emerald
Cordulia aenea (Linnaeus)

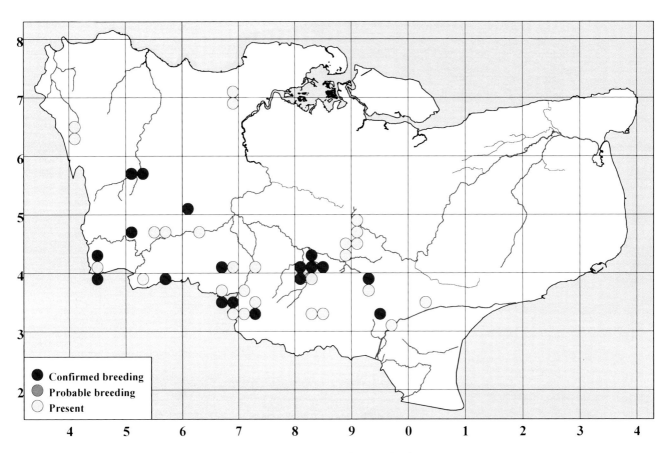

Confirmed breeding
Probable breeding
Present

Description

This is a medium sized dragonfly with dark metallic emerald green thorax and abdomen. The abdomen has bronze highlights, especially in the male, whose abdomen is club shaped and noticeably 'waisted' at segment 3. The thorax is very downy due to a dense coat of brownish hairs, and the eyes of the mature adult are deep green. The wings are clear with a deep amber patch at the root of each. The pterostigmata are rectangular and blackish. This species prefers to settle away from the water, often hanging vertically from branches to consume its prey.

Habitat

Cordulia aenea requires still waterbodies, such as lakes, ponds and canals in woodland areas. There are usually trees shading at least part of the margin and often little or no submerged aquatic plants.

Breeding

The female deposits her eggs whilst in flight by tapping her abdomen into the water and allowing the eggs to be washed off. This usually takes place in

Downy Emerald, male (Nick Donnithorne)

unshaded areas and where aquatic vegetation is not dense. The larvae live in coarse leaf litter, avoiding plants, decomposed leaves and silt. When development is complete the larvae climb stems in the water close to the edge, or they may sometimes crawl a distance from the water to emerge as adults on rushes or other suitable plant supports. Exuviae have been found hanging from the tips of the branches of young pine trees two metres from the ground.

Distribution

Cordulia aenea has a localised scattered distribution mainly south of the River Thames with some further populations in Scotland and Wales. In Kent there are scattered populations mainly confined to the south-western quarter of the county. Proven breeding has been established at several locations such as at Marshall's Lake, Bedgebury Pinetum at Chart Hills Golf Course near Biddenden and at Sissinghurst Castle Lakes. Longfield (1949a) mentions *C. aenea* as being present in V. C. 15 and 16, and specifically that it was common on the Keston and Chistlehurst ponds (Longfield, 1949b).

Status

Regionally important in VC15 and VC16.

Similar species

Somatochlora metallica - has an extensive yellow band across the frons.

Northern Emerald, *Somatochlora arctica* - is only found in Scotland.

Flight period in Kent

Late April to late July.

Brilliant Emerald
Somatochlora metallica (Vander Linden)

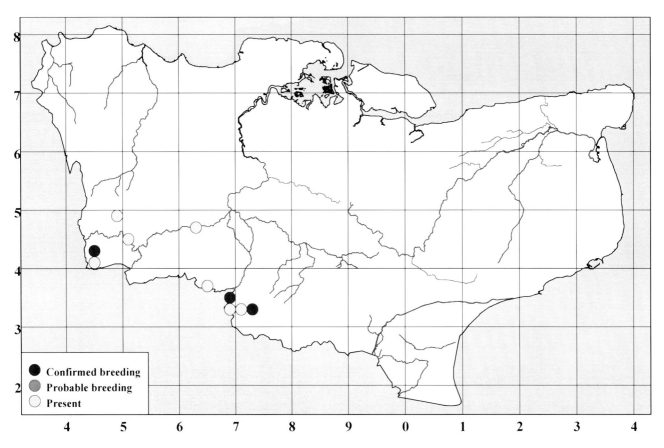

Brilliant Emerald, male (Mike Thurner)

Description

This species is a medium sized dragonfly with a bright metallic green thorax and abdomen. The abdomen lacks the bronze highlights of *Cordulia aenea* and appears brighter in colour. The eyes are deep green and there is a band of yellow across the frons. The wings are lightly suffused with an amber colouration and the female has deep amber patches nearer to the front margins at the base of the wings. The pterostigmata are long and yellow-brown in colour. The female's vulvar scale, which hangs down at a 90° angle to the abdomen, can often be seen in flight.

Habitat

Somatochlora metallica inhabits still waterbodies, such as large ponds, lakes, and canals and occasionally slow flowing rivers. There are usually trees shading at least part of the margin and little aquatic vegetation. Woodland, often coniferous, is necessary for feeding as this species rarely, if ever, feeds by the water (Brooks, 1997).

Breeding

The female, using her vulvar scale, oviposits among exposed tree roots or in the soil in shaded areas. The larvae live in the leaf litter avoiding areas clear of this debris. After two or three years the larvae leave the water to emerge on rushes or bushes at the water margin or on the trunks of trees where exuviae may be found in the bark fissures as much as two metres above the ground.

Distribution

There are two separate populations of *S. metallica* in Britain. One is in Scotland, the other in south-eastern England. It was thought to be rare or extinct in Kent (Longfield, 1949a), although it had been recorded by Stephens (Longfield, 1949b). Longfield suggested that although it requires a similar habitat to *Cordulia aenea*, it probably cannot successfully compete with this more dominant species. Latest records indicate that this species is confined to just a few locations in the south-western part of the county. Breeding has been recorded at Marshalls Lake, Bedgebury Pinetum, where exuviae were found by the authors in July 1996, and more recently at Colliers Pond, Scotney and at Christmas Mill.

Status

Nationally important

Similar species

Cordulia aenea - has no yellow band on the frons.

Northern Emerald, *S. arctica* - is only found in Scotland.

Flight period in Kent

Mid June to late July.

Freshly emerged Brilliant Emerald, with exuviae
(Nick Donnithorne)

Four-spotted Chaser

Libellula quadrimaculata Linnaeus

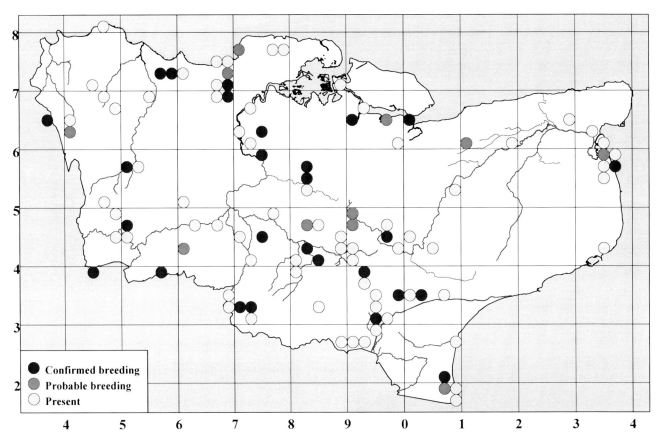

Description

This is a medium sized dragonfly of relatively heavy build. The dark olive brown colour of the thorax and of the abdominal segments 1-6 is broken into a 'crazy paving' pattern. Segments 7-10 are black. Each segment from 4-8 or 9 has an elongated yellow spot on each side. Immature specimens are orange in colour before becoming the typical dark olive brown. The eyes are dark brown. All four wings have a dark brown spot at the node, giving this species its name. These spots are linked to the wing base by an amber costal band. There is also a large dark brown patch at the base of the hind wings. The pterostigmata are long and dark brown with a tiny smudge immediately below. Sometimes this smudge is larger, reaching the hind edge of the wing, and the spots are elongated with the amber band extending to the pterostigma. Individuals with these markings are known as form *praenubila* (Newman). This form has been observed at Sissinghurst Castle lakes.

Four-spotted Chaser, male (Mike Thurner)

Habitat

Libellula quadrimaculata is found at many different types of still and slow-flowing waterbodies such as bogs, ponds, lakes, canals and streams. These may or may not be acidic, or may even be brackish.

Breeding

The female lays her eggs into the water, while flying, by dipping the tip of her abdomen into the surface usually near submerged plants. The male is often patrolling nearby ready to chase off rival males. Larval development may take two years. Exuviae may be found in suitable vegetation, often rushes, at the margin or several metres away from the water. In bogs they emerge on sphagnum moss, heather or bog myrtle.

Distribution

Libellula quadrimaculata is widespread over much of Britain, though scarce in the north-east. In lowland England it has a preference for acid water. In Kent, it has a widely scattered distribution. On the 6th and 7th June 1889 thousands of this species passed Dover flying against a northeast wind and on the 10th June 1900 "vast numbers arrived" at Margate (Dannreuther, 1941). Longfield (1949b) reported that it had "not yet been seen in north-west Kent, although moderately frequent along the south-east coast from migrations and no doubt also breeding in suitable places".

Status

Regionally important in VC15 and VC16

Similar species

Libellula depressa - especially the over mature female.

Libellula fulva - especially the over mature female. Also the normal female colour and that of the immature male *L. fulva* resemble the immature *L. quadrimaculata.*

Flight period in Kent

Late April to early August.

Scarce Chaser
Libellula fulva (Müller)

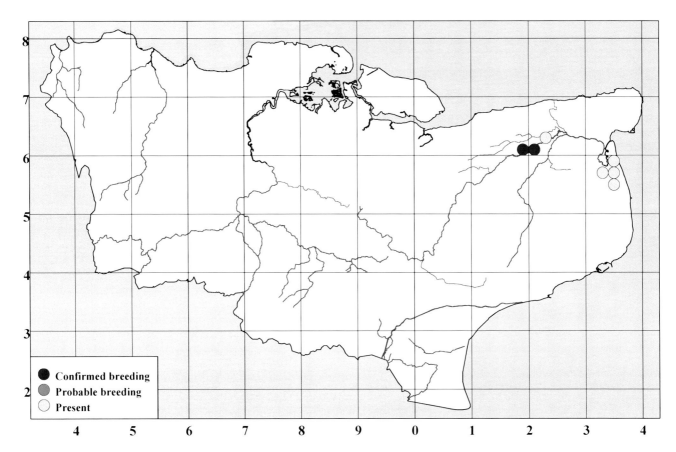

Description

This is a medium sized dragonfly of relatively heavy build. The sexes are so different as to appear as two species. The male has a dark grey thorax and the abdomen is powder blue on segments 3-7, and black on segments 1,2 and 8-10. There is a dark brown patch at the base of the hind wing and the pterostigmata are dark brown also. There may be a small dark brown spot at the wing tip. The eyes are dark greyish blue. The female has a grey-brown thorax and bright orange abdomen with a black dorsal stripe, which is thin at first and becomes broader from about segments 5-10. The abdomen of the immature male is of similar colour and pattern. The wings of the female have dark brown patches at the base and have amber costal bands reaching just beyond the nodes. The pterostigmata and wing tip spots are dark brown. The eyes are grey-brown.

Habitat

Libellula fulva is found by slow to moderate flowing rivers, lakes and gravel pits with ample emergent and marginal vegetation. It likes sunny sheltered areas and avoids heavy shade.

Scarce Chaser, female

Scarce Chaser, male

Breeding

With the male usually nearby, the female oviposits in open water, at the edge where the flow is slower, by continually dipping the tip of her abdomen into the surface whilst in flight. Usually after two years, when the larvae are ready to emerge, they climb a suitable support over the water and close to the waters edge. Common reed *Phragmites australis* (Cav.) Trin. ex Steudel, Sweet-grass (*Glyceria* spp.) and sedges (*Carex* spp.) are some of the plants on which exuviae have been found.

Distribution

This rare species occurs in just a few scattered locations in southern England where it may be locally abundant. In Kent a female was observed at Kingsdown near Deal in 1881 by C. G. Hall (Lucas, 1900) and a single male was seen at Sandwich on August 22nd 1898 by W. J. Lucas. "Stephens gives it (1836) for Deptford and Bermondsey and Evans (1845) for Herne Bay, but it has not been found in Kent nearer than the Sandwich and Deal districts in later years" (Longfield, 1949b). It was rediscovered in Kent by workers from the Sandwich Bay Bird

Observatory in 1985 at North Stream near Sandwich and has occasionally been seen in subsequent years. On 24th May 1998 it was discovered along the River Stour at Westbere by Ian Hodgson of Canterbury. Exuviae were found along the river and adjacent gravel pit lake by the authors on 29th May 1998 when 15-20 adults were also recorded. This colony continues to thrive at Westbere, where exuviae have since been found on a second lake.

Status

Nationally important

Similar species

Libellula quadrimaculata - immature are similar to female Libellula fulva.

Orthetrum cancellatum - the males are similar.

Orthetrum coerulescens - the males are similar.

Flight period in Kent

Early May to mid July.

Broad-bodied Chaser

Libellula depressa Linnaeus

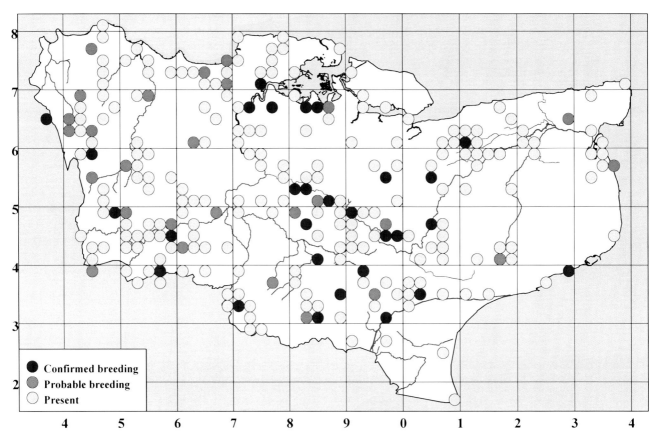

Confirmed breeding
Probable breeding
Present

Description

This species is medium sized and heavily built with a very broad or depressed abdomen. The male thorax is dark brown with pale blue green antehumeral stripes, and the abdomen is sky blue with a pair of yellow lateral spots on each segment. The female abdomen, which is broader than the males, is yellow-brown with larger yellow lateral spots. Immature males are similar in colour to females. The wings of both sexes have a dark brown patch at the base, and the pterostigmata are long, narrow and black. The eyes are dark brown. This species often returns repeatedly to the same perch to watch for prey or to consume it.

Habitat

Libellula depressa is one of the first species to colonise a newly created pond, and favours still waterbodies such as ponds and ditches. Its numbers may decrease if aquatic and marginal vegetation become too dense, but some clearance work will often encourage their return.

Breeding

When ovipositing, the female, whilst in flight, frequently dips the tip of her abdomen into the water where there is submerged vegetation. The eggs are washed off and adhere to the plants beneath. The male is usually in attendance to chase off rivals, but females are also often seen egg laying alone. Larval development can be completed in one year, but may take longer. Exuviae may be found amongst a variety of marginal vegetation, but rushes (*Juncus* sp.) are preferred when available.

Distribution

Libellula depressa is widespread and often common in much of southern Britain but scarcer in the Midlands and East Anglia. It is absent from the north. In Kent, it is widespread but usually seen in low numbers. There is an early record for this species from Folkestone Warren in June 1879 (Ullyett, 1888).

Broad-bodied Chaser, male (top) and female (bottom) (lower photograph: Greg Hitchcock)

Similar species

Libellula fulva - the males are similar.

Libellula quadrimaculata - is similar to the female and immature male of *L. depressa*.

Orthetrum cancellatum - the males are similar.

Flight period in Kent

Mid April to mid September.

73

Black-tailed Skimmer

Orthetrum cancellatum (Linnaeus)

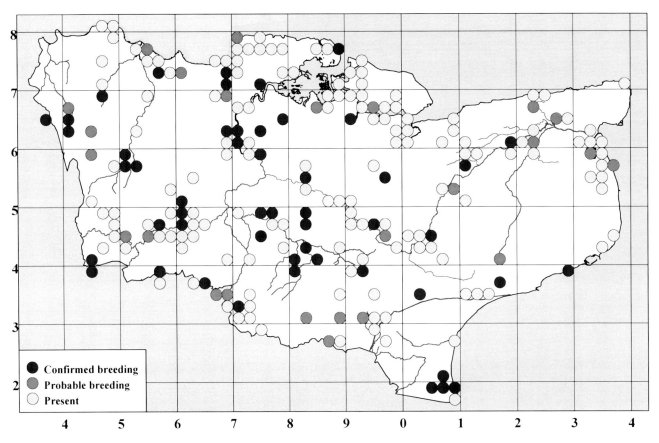

Description

This is a fairly large dragonfly with a grey brown thorax. The male's abdomen has a sky blue pruinescence except for segments 1 and 2, which are brown, and 8-10, which are black. Segments 3-8 have elongated yellow lateral spots. The female's abdomen is yellow with a pair of black longitudinal stripes on each segment. The immature male is of similar colour and pattern. When over mature, her abdomen gains a dark grey-blue pruinescence. The wings of both sexes are clear with dark brown or black pterostigmata, and the eyes are dark grey-brown. This species frequently rests on bare soil or stones.

Habitat

Orthetrum cancellatum favours lakes and gravel pits, sometimes with part of the margin devoid of vegetation. Numbers may decrease if vegetation becomes too dense. Slow rivers, canals, marshland dykes and small ponds also sometimes support populations.

Black-tailed Skimmer, female

Breeding

The male usually guards the female while she oviposits. She does this in flight by dipping her abdomen down into the water. When ready to emerge the larvae leave the water and climb amongst the marginal rushes and plants. Exuviae may be found in almost any type of herbage, including grasses. This may be at the edge, or some distance away from the water. The authors once found an exuviae hanging from a heather clump 10 metres from the water.

Distribution

Orthetrum cancellatum is fairly common and widespread in southern England and south Wales. Stephens reported it to be "not uncommon" in west Kent at Crayford and Dartford marshes, but later it was said to be "only found in East Kent, nearer to the sea" (Longfield, 1949b). It is now common and widespread, especially as the county has numerous fishing lakes and flooded gravel pits, such as at Dungeness, Chilham Lakes and Westbere.

Similar species

Libellula fulva - males only

Flight period in Kent

Mid May to mid September

Black-tailed Skimmer, male

Keeled Skimmer
Orthetrum coerulescens (Fabricius)

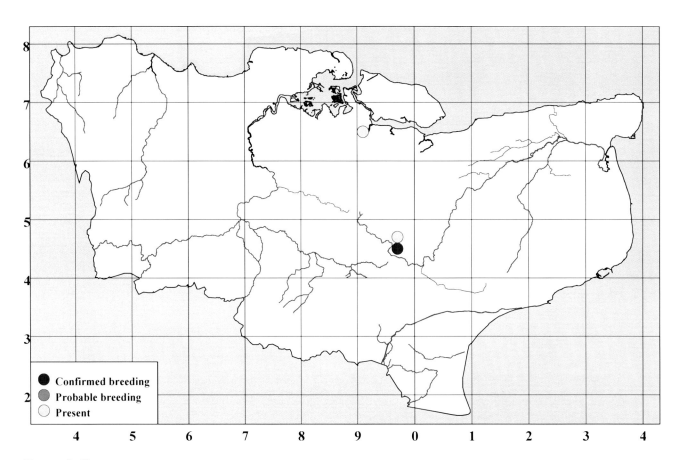

Confirmed breeding
Probable breeding
Present

Description

This species is smaller than *O. cancellatum* and has a dark grey thorax with pale antehumeral stripes. The male has a pale powder blue pruinescence on the abdomen with the exception of segment 1 which is black. In older males a grey-blue pruinescence may also cover the thorax. Immature males are yellow with heavy black borders on the abdominal segments. The female's abdomen is straw yellow making her look much like a darter (*Sympetrum* sp.). In over mature specimens her abdomen becomes dark blue-grey. The wings of the female are lightly suffused yellow. The pterostigmata are long and yellow, and the eyes are dark grey.

Habitat

This species occurs in acid bog pools, small shallow streams and ditches where *Sphagnum* mosses grow and there is little competition with other species of dragonfly.

Breeding

While the male keeps a protective watch from a nearby perch or in flight, the female, using her abdomen, flicks small drops of water with some eggs onto nearby vegetation. Sometimes she may be seen doing this alone without the male in attendance. Larval development takes about two years. After emergence the exuviae may be found clinging to short vegetation typical of acid bogs, and often close to the ground.

Distribution

Orthetrum coerulescens is locally common in heath and moorland bogs in southern and south-western England, also Wales and parts of northern England and western Scotland. In Kent, it was only known from the main bog at Hothfield Common. In 1999 exuviae were abundant, and in one small area 50 or more were found by the authors in a few minutes of searching. On 25.09.03 two adult specimens were seen by Dr. Phil Smith near Kemsley. One was at a fishing lake at Church Marshes and the other was at Kemsley Drain. A photograph was taken of one adult which was later identified by Tim Beynon of the British Dragonfly Society. However, the authors could find no suitable breeding habitat at this location and there have been no further sightings of this species from there. This species was recorded at Chattenden in the past by H. J. Turner (Lucas, 1900).

Status

Regionally important in VC15

Similar species

Orthetrum cancellatum - but this species is larger and has a black tipped abdomen.

Flight period in Kent

Late May to late August.

Keeled Skimmer, male

Common Darter

Sympetrum striolatum (Charpentier)

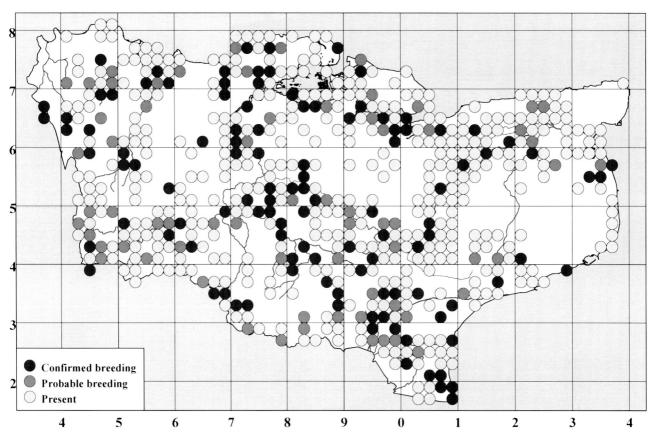

- ● Confirmed breeding
- ● Probable breeding
- ○ Present

Description

This is a small dragonfly with a brown thorax, which in the male has small pale antehumeral stripes and broad yellow lateral bands. The male's abdomen is orange-red with a small central black marking on segments 8 and 9. Females have a yellow abdomen with the same black markings on segments 8 and 9. Immature males are yellow and resemble the females. The clear wings of both sexes have rectangular brown pterostigmata. The eyes are brown and the legs are striped longitudinally black and yellow. *Sympetrum striolatum* often rest on bare ground in cooler conditions to take advantage of any residual warmth.

Older females may take on a reddish colouration similar to the male. It is often the last dragonfly to be seen flying even in late November.

Habitat

This species may be found in almost any type of waterbody such as lakes, ponds, ditches and rivers with aquatic and marginal vegetation. It is also often seen in woodland rides and clearings in large numbers.

Breeding

After copulation, with the male often still attached *in tandem*, the female will oviposit by repeatedly dipping her abdomen into the water near submerged vegetation. This may take place in one spot for a while before moving on to another. Sometimes the female will oviposit alone. Eggs laid early in the season hatch quickly and take one year to complete larval development. Eggs laid late in the season hatch the following spring. Emergence takes place on a variety of plants both out in the water and at the margin. Exuviae are frequently found in large numbers.

Distribution

Sympetrum striolatum is common and widespread throughout much of England and Wales and also parts of Scotland, though not so common there. In Kent it is common and widespread and often abundant. During a visit to Smallmans Wood near Hamstreet in August 1990, the authors encountered at least one hundred of this species feeding in the woodland rides and clearings.

Similar species

Sympetrum sanguineum - which has a club-shaped abdomen, black legs and no antehumeral stripes.

Flight period in Kent

Early June to early December.

Vagrant Darter

Sympetrum vulgatum (Linnaeus)

Description

This species is a small dragonfly which has a brown thorax with pale antehumeral stripes. Unlike the similar *S. striolatum*, it has no broad yellow lateral bands. The male's abdomen is orange-red with much reduced black markings on segments 8 and 9. The immature male and the female have a straw-yellow abdomen with similar black markings. The female's vulvar scale is held at 90° to the abdomen. The clear wings sometimes have very small yellow basal spots. The red-brown pterostigmata have a black border and the eyes are brown.

Status

This is a rare migrant to Britain from central and north-eastern Europe, which breeds in still waterbodies. In Kent there are records from the Bird Observatory at Dungeness (TR01Y) during this survey. These are from 1993 when three were trapped on the 3rd August, one on the 5th, 6th, 8th and finally a single individual on the 9th. Due to the difficulty in separating this species from *S. striolatum*, only these records of *S. vulgatum* which had been trapped and photographed in the hand were accepted (Attridge, 1996). There are also historical records for the county (Longfield, 1949a). One such record states that a female was caught at Keston in 1925 (Longfield, 1949b). Another record states that this species was seen in Kent in August-September 1938 (Dannreuther, 1939).

Similar species

Sympetrum striolatum - these two species are very difficult to separate. If examined in the hand, the extent of the black marking along the top and sides of the frons can be established.

Vagrant Darter, male (Danny Chapman)

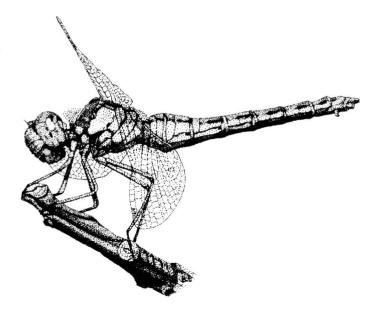

Red-veined Darter

Sympetrum fonscolombii (Sélys-Longchamps)

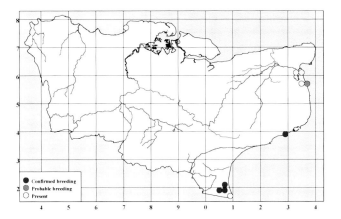

Description

This is a medium sized darter. The male has a red-brown thorax and a bright red abdomen with black markings on segments 8 and 9, and the eyes are reddish. Females and immature males have a yellow thorax and abdomen with the same black markings on segments 8 and 9, and the eyes are brown. The red wing venation, which gives the species it's name, is found only in the male. The female's wings have yellow veins. Both sexes have yellow basal wing patches, and pale yellow pterostigmata with black borders.

Habitat

Sympetrum fonscolombii breeds in shallow ponds, lakes and ditches, and will tolerate brackish conditions.

Breeding

As with most darters, the female can be found ovipositing with the male attached *in tandem* or on her own. The larvae develop extremely rapidly and two generations of adults can be produced in one year in southern Europe. Exuviae in one Kent location were found on the stems of Common Reed *Phragmites australis* a short way from the water. The exuviae can be separated from other darter species by the absence of dorsal spines and extremely short lateral spines on segment 9.

Distribution

Sympetrum fonscolombii is found throughout southern Europe with occasional migrations northwards. It is a frequent migrant to Britain most years in small numbers. One author states that this species was recorded in Kent in the 19th century (Dannreuther, 1939). Another more specific early Kent record is of one male at Deal in 1881 (Lucas, 1900), and pre 1910 in V.C. 15 (Longfield, 1949a). On the RSPB. reserve at Dungeness in June 1996 at least ten adults were seen, including a pair ovipositing. A teneral female was seen on 18th August 1996 which may have been bred from the spring immigration or, less likely, from an undetected presence in 1995. None was seen on the reserve in 1997 but ovipositing was again observed in 1998 and one exuviae was found. Mating was also observed in 1999 (pers. comm. RSPB Dungeness), and probably as a result of this, and not migration, at least 49 adults were counted in 2000. Two more exuviae were found in 2004. The colony continues to survive though numbers in more recent years have been disappointingly low. In 1999 two exuviae were found by the authors at Samphire Hoe near Dover, and single exuviae was found there in 2002. Small numbers of adults continue to be seen at Samphire Hoe each year.

Similar species

Most red darters are similar but the red or yellow veins which are most noticeable near the leading edge of the basal half of each wing and the pterostigmata are diagnostic of *S. fonscolombii*.

Flight period in Kent

Late May to late August.

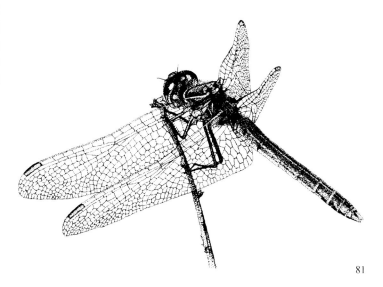

Yellow-winged Darter

Sympetrum flaveolum (Linnaeus)

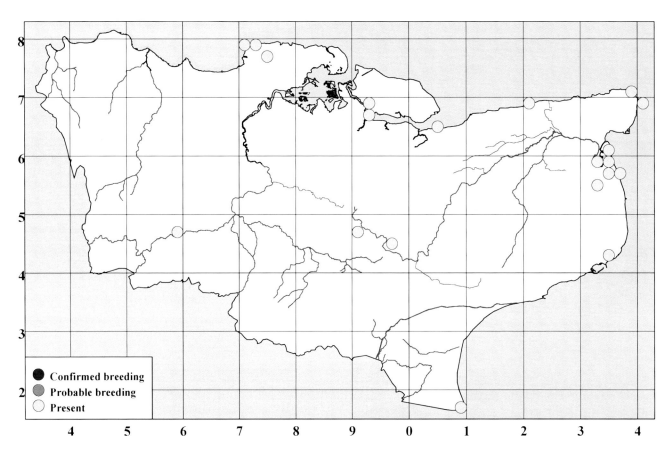

Description

This is a small dragonfly. The male has a red-brown thorax, and the orange-red abdomen has black markings on segments 8 and 9. The abdomen of immature males and adult females is straw-yellow with these same black markings. The most obvious identification feature is the large basal yellow patch on the wings, which may cover as much as one third of the wing area. The pterostigmata are red-brown bordered with black, and the eyes are brown.

Habitat

This species may be found at bog and marsh pools, ponds and lakes with aquatic and marginal vegetation.

Breeding

The female is usually *in tandem* with the male during egg laying. She oviposits in vegetation at the shallow edge of the water or sometimes in damp soil which is covered by water during autumn and winter. Breeding has been recorded in England but not sustained. No breeding has been recorded in Kent.

Yellow-winged Darter, female (Danny Chapman)

Distribution

Sympetrum flaveolum is found throughout most of Europe, except in the southern parts of Spain, Portugal, Italy and Greece, and the northern parts of Sweden, Norway and Finland. In Britain it is an occasional migrant which in some years occurs in large numbers, especially in 1926, 1945, 1955 and 1995 (Brooks, 1997) In Kent it has been recorded from 20 scattered tetrads, many sightings being of one individual only, while others are of 30, 40 or even 170 individuals (Attridge, 1995, 1996) at Dungeness, possibly due to its proximity to the continent. The bulk of the records came from 1995 when a huge influx of this species 'invaded' Britain on strong, warm easterly winds. An unexpected sighting was on 7th October 1995 at Hothfield Common during a Kent Field Club meeting (Brook, 1996). An early record for this species is from "Abbey Wood and at Swanscombe" in west Kent (Longfield, 1949b).

Similar species

Most other red darters are similar, but none have the extensive yellow wing patches of *S. flaveolum*.

Flight period in Kent

Migration records cover the months June to October.

Ruddy Darter

Sympetrum sanguineum (Müller)

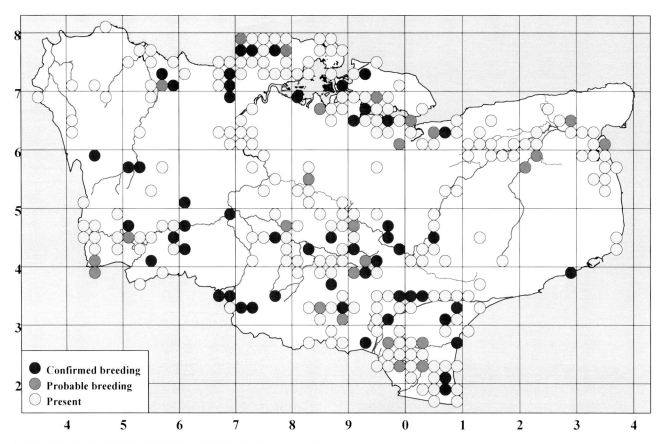

Ruddy Darter, female

Description

This is a small dragonfly with a red-brown thorax without antehumeral stripes. The male's abdomen is a deep crimson colour and club shaped. Immature males are similar in colour to adult females which have a yellow abdomen. Both sexes have black markings on segments 8 and 9. Each wing has a small yellow basal spot. The pterostigmata and the eyes are brown. Unlike other red darters this species has black legs. Older females may become reddish.

Habitat

Sympetrum sanguineum favours shallow lakes, ponds, ditches, bog pools and canals with plenty of aquatic vegetation. Woodland ponds are frequently used.

Breeding

With the male still attached *in tandem*, the female dips her abdomen into the water to wash off the eggs. This may take place among aquatic or marginal plants and sometimes in damp soil near the waters edge. The authors have noticed them ovipositing among grasses

a metre or more from the water. In Kent, it has also been observed ovipositing in damp leaf litter in an oak woodland where the area fills with water during the autumn. Late season eggs may not hatch until the following spring. Larvae climb aquatic or marginal plants for emergence. The exuviae are difficult to distinguish from those of *S. striolatum* whose lateral spines on segment 9 are generally longer than those of *S. sanguineum*. However, often due to some variation of the length of these spines, this is not always a reliable aid to identification.

Distribution

This species is most common in England south of a line from The Wash to west Dorset and also around the Bristol Channel area. Further populations occur north of this line up to the Rivers Humber and Ribble, and in South Wales. Frequent migrations from Europe boost the population. In Kent it is quite common and widespread, especially at ponds in or near woodland, such as Burnt Oak Wood and at Chiddingstone. Earlier records for this species in Kent were at Deal and Dover District by C. G. Hall, Gravesend by H. J. Turner and Sandwich by W. J. Lucas (Lucas, 1900). Breeding was recorded in V. C. 15 and 16 (Longfield, 1949a).

Similar species

Sympetrum striolatum - this species is a little larger, has black and yellow legs and the male has an orange-red abdomen.

Flight period in Kent

Early June to mid October.

Black Darter

Sympetrum danae (Sulzer)

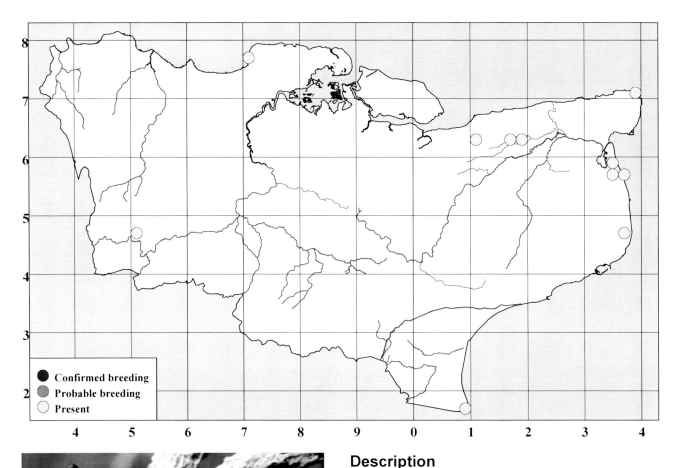

Black Darter, male

Description

This small dragonfly is the only black species of darter in Europe. The mature male has a black thorax and abdomen. There may be some yellow markings on the sides of segments 1-4 and paired yellow markings on segments 7 and 8. The eyes are black. Females and immature males have a yellowish thorax with a characteristic black triangle on top. The yellow abdomen has black markings on segments 7 and 8. The eyes are brown and there are small yellow basal patches on the female's wings. In both sexes the legs and pterostigmata are black. Another important feature of this species is the presence of three yellow spots on a central black band on the side of the thorax.

Habitat

Sympetrum danae may be found in shallow acidic heathland and moorland bog pools with abundant aquatic vegetation.

Breeding

Oviposition usually takes place with the pair *in tandem* at first, and then the female may continue

alone. *Sphagnum* moss, well known for its water retention ability, will protect the larvae should the pool dry up temporarily. The larvae develop very rapidly and, after only two months or so, they climb emergent vegetation in or near the waters edge. The exuviae are very small and can be distinguished from the other *Sympetrum* species in this country by the very small lateral spines on segment 9.

Distribution

This species is widespread throughout much of northern Europe and northern Britain. It is also found in the heathlands and moorlands of southern England. *Sympetrum danae* was once found sparsely in Kent (Longfield, 1949a). One adult male was found on the River Darent on the 17th August 1947 (Longfield, 1949b). It is not now recorded in the county except as a migrant from the continent. The most recent records coincided with the large influx from Europe of *S. flaveolum* and other species which arrived during a period of strong easterly winds in 1995.

Similar species

Most female and immature male darters are similar to the same of this species. However the small size, black triangle on the top of the thorax and the 3 yellow spots on the sides, separate *S. danae* from the others. The black legs also separate it from all but *S. sanguineum*.

Flight period in Kent

Late July to mid September.

Globe Skimmer
Pantala flavescens (Fabricius)

Description

This rare migrant is a large dragonfly quite unlike any other libellulid species found in Britain. The thorax is yellow-brown with no antehumeral stripes. The abdomen of the male is yellowish with reddish markings on each segment. There is also a black dorsal stripe which becomes broader forming triangular markings on segments 8-10. The female is similar but has a more olive abdomen. The wings of the male may have a light brown smudge at the tip. There is also an amber patch on the broad basal lobe of the hind wing especially in the female. The red-brown pterostigmata are longer on the fore wing than on the hind wing. The eyes are brown and the legs are dark with a yellow longitudinal stripe.

Distribution

This global tropical migrant is occasionally seen in Spain and southern France. A specimen was taken at Horning, Norfolk by Sparshall (Lucas, 1900). It has been recorded in Britain only four times, the most recent record being from Kent in 1989 when a single individual was found near a transcontinental lorry depot, and presumed to be an accidental introduction (Brooks, 1997, Parr, 2001). The specimen was passed to Mr. Alan Davies, an odonatologist, and resides in his collection.

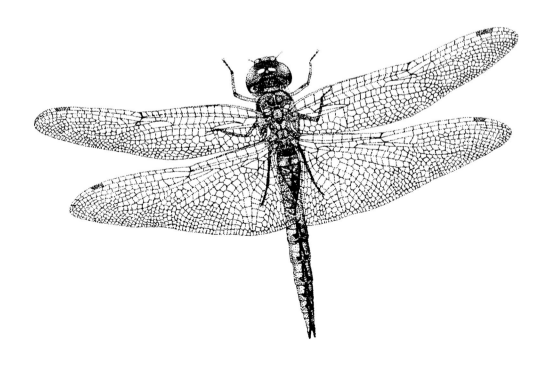

Species no longer recorded in Kent

Common Hawker

Aeshna juncea (Linnaeus)

This species is very similar to *A. mixta* but both sexes lack the yellow triangular marking on segment 2, and the anal triangle of the male's hind wings have only two cells not three, as in *A. mixta*. It is widespread in northern and western Britain and also the heathlands of the south and east. Early records indicate that *A. juncea* has been seen in Kent presumably after migration (Longfield, 1949a). It was recorded that this species was found at Ham Street on 12th August 1919 by A. E. Tonge (Dannreuther, 1939). In recent years there have been occasional, but as yet unconfirmed, reports of this species in the county.

This dragonfly is usually on the wing from late June to early October.

Common Hawker, freshly emerged female (Nick Donnithorne)

Southern Migrant Hawker

Aeshna affinis Vander Linden 1823

This rare migrant species is very similar to *A. mixta*, but all abdominal markings including those on segment 2 are deep blue. The first confirmed British record for this dragonfly was on 5th August 1952 when one was taken by W. E. Dyson on Romney Marsh (Merritt, Moore & Eversham, 1996).

Southern Migrant Hawker (Mike Thurner)

Vagrant Emperor

Hemianax ephippiger (Burmeister)

This species is similar to *Anax parthenope* but has a yellow brown abdomen and brown eyes. The blue on the abdomen is confined to segment 2 only and is deeper in colour.

There is one historical record only for *H. ephippiger* in Kent. A single male specimen was picked up from the pavement in the centre of Tunbridge Wells one evening, late in October 1967, by the son of D. A. L. Davies. It is now in D. A. L. Davies' collection (see record cards in the Kent Biological Archives and Records Centre, Maidstone Museum.) Most individuals arrive during the late summer and autumn, and most have been recorded from the south of Britain but a few also from the north, including one from the Shetland Isles (Brooks, 1997).

Common Club-tail

Gomphus vulgatissimus (Linnaeus)

The male is a medium-sized black and green dragonfly with a pronounced club shaped abdomen. The male has bright yellow markings on the sides of abdominal segments 7 to 9. The female and the immature male are black and yellow. It is on the wing from mid May to the end of June.

In Britain, it is local and exuviae are often found in very large numbers. It is a riverine species found in parts of south-west Wales, along the area of the Welsh border, on stretches of the River Thames, and also on the River Arun in Sussex. Lucas states that "Evans on Stephen's authority gives 'round London' and 'in Kent'" (Lucas, 1900). There are also records for this species at Dartford (Bath, 1890) and the River Darent in June 1939 (Longfield, 1949b). Longfield records that she had no idea how it came to be in that locality and that it had not been proved breeding there.

Large White-faced Darter

Leucorrhinia pectoralis (Charpentier)

As with *Pantala flavescens,* this species must be seen as an accidental introduction when according to Lucas (1900) one adult was recorded near Sheerness in June 1859. It may not have made landfall by itself, but was probably taken on board a fishing boat (McLachlan, 1884). In January 1860, it was exhibited at a meeting of the Entomological Society of London.

This species is similar to the White-faced Darter *Leucorrhinia dubia,* a resident British species, which is now not found south of Staffordshire. *L.pectoralis* is blackish with red brown antehumeral stripes and red brown patches covering most of each abdominal segments 1-6. Segment 7 has a large yellow spot. Segments 5-7 are expanded to form a club shape. The face is white.

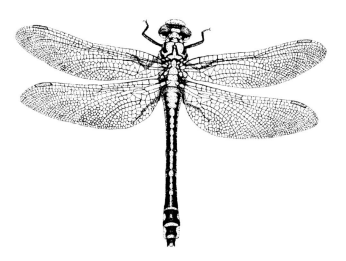

Common Club-tail, female

Flight periods

Flight periods from Kent based on records received during the survey period from 1980-2008

Adapted from Hammond, 1983:-

■ Optimum flight times

▨ Earlier and later dates when species may be teneral or very adult

▫ Exceptionally early or late dates, often influenced by weather conditions

Species	March	April	May	June	July	Aug.	Sept.	Oct.	Nov.
Calopteryx virgo									
Calopteryx splendens									
Lestes sponsa									
Lestes dryas									
Platycnemis pennipes									
Pyrrhosoma nymphula									
Erythromma najas									
Erythromma viridulum									
Coenagrion puella									
Coenagrion pulchellum									
Enallagma cyathigerum									
Ischnura elegans									
Aeshna mixta									
Aeshna cyanea									

Species	March	April	May	June	July	Aug.	Sept.	Oct.	Nov.
Aeshna grandis				░	▓	▓	▓░	░	
Anax imperator			░	▓	▓	░			
Anax parthenope									
Brachytron pratense			░	▓	░				
Cordulegaster boltonii				░	▓	▓	░	░	
Cordulia aenea			░	▓	░				
Somatochlora metallica				░	▓	░			
Libellula quadrimaculata			░	▓	▓	░			
Libellula fulva			░	▓	░				
Libellula depressa			░	▓	▓	░			
Orthetrum cancellatum			░	▓	▓	░			
Orthetrum coerulescens				░	▓	▓	░		
Sympetrum striolatum				░	░	▓	▓	░	░
Sympetrum vulgatum					▓	▓	▓		
Sympetrum fonscolombii				░	▓	▓	░		
Sympetrum flaveolum					▓	▓	░		
Sympetrum sanguineum				░	▓	▓	░		
Sympetrum danae				░	▓	▓	▓	░	

First and last dates of adult sightings

This list gives the first and last recorded annual dates of adult sightings in Kent. In each case, the year in which the sighting was made is also given.

Species	First Date	Last Date
Calopteryx virgo	10 May 2008	30 August 1991
Calopteryx splendens	4 May 2003	14 September 2007
Lestes sponsa	11 May 2007	27 September 1980
Lestes dryas	24 May 2007	16 August 2003
Platycnemis pennipes	30 April 2007	23 September 2007
Pyrrhosoma nymphula	12 April 2007	26 August 1999
Erythromma najas	2 May 2007	21 September 2002
Erythromma viridulum	24 June 2003	21 September 2006
Coenagrion puella	28 April 2007	10 September 2000
Coenagrion pulchellum	7 May 2008	15 August 2002
Enallagma cyathigerum	27 April 1986	6 October 1994
Ischnura elegans	16 April 2007	1 October 1996
Aeshna mixta	1 July 2006	19 November 2006
Aeshna cyanea	29 May 1997	14 November 2008
Aeshna grandis	25 May 1999	6 October 1991
Anax imperator	27 April 2007	23 September 2006
Anax parthenope	17 June 2006	23 August 2000
Brachytron pratense	15 April 2007	11 July 1991
Cordulegaster boltonii	21 May 2002	31 August 2005
Cordulia aenea	26 April 2007	28 July 1999
Somatochlora metallica	16 June 1999	24 July 2006
Libellula quadrimaculata	30 April 2007	5 August 2003
Libellula fulva	3 May 2000	22 July 2008
Libellula depressa	17 April 2007	25 September 1997
Orthetrum cancellatum	20 May 1995	12 September 1991
Orthetrum coerulescens	29 May 1998	25 August 2002
Sympetrum striolatum	2 June 2005	9 December 2006
Sympetrum vulgatum	3 August 1995	8 August 1995
Sympetrum fonscolombii	31 May 2004	25 August 2006
Sympetrum flaveolum	19 June 2007	7 October 1995
Sympetrum sanguineum	15 May 2000	30 October 2004
Sympetrum danae	21 July 2006	21 September 1996

Exuviae and their role in dragonfly recording

The collecting and identifying of exuviae has played a crucial role in the recording of dragonflies in Kent. Firstly, it provides a positive identification of a species and secondly, it establishes that the species is breeding at the site where the exuviae were found.

Exuviae of Downy Emerald (Nick Donnithorne)

The collecting of exuviae is not dependent on the weather. Even when the weather is cold and wet, records can be obtained. New sites have been discovered for species when exuviae have been found where the adults have not been recorded. The lack of adult sightings at such places can possibly be due to inclement weather, or shortly after emergence when the teneral dragonflies go away from the water to mature, or sometimes later in the season when the adult is no longer on the wing. One very notable example is that of the exuviae of the Willow Emerald Damselfly, *Lestes viridis* which was collected from Cliffe Marshes in 1992 (Brook & Brook 2003). No adults of this species have ever been recorded in Kent, so with the discovery of the exuviae, this was solid proof of its existence there at that time. Unfortunately, it has not been recorded in Kent since.

The collecting and identifying of exuviae can be of great value in establishing the validity of adult records. Exuviae provide proof of breeding which is often crucial in site evaluation and protection, and regular collecting of exuviae can provide a cumulative record of emergences (Merritt, Moore & Eversham, 1996). At one site in Kent, 52 exuviae of the Downy Emerald Dragonfly, *Cordulia aenea* were collected over a three-month period although only up to three or four adults were seen at any one visit. At another site a total of 346 exuviae of the Southern Hawker, *Aeshna cyanea,* 477 exuviae of the Ruddy Darter, *Sympetrum sanguineum* and 285 exuviae of

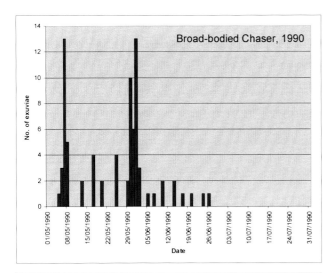

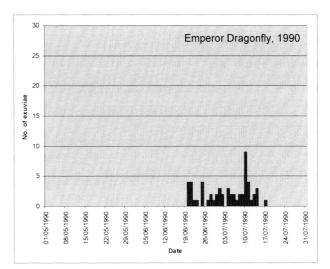

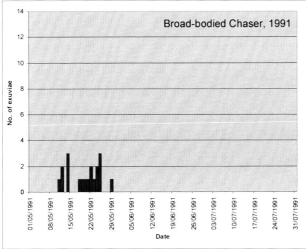

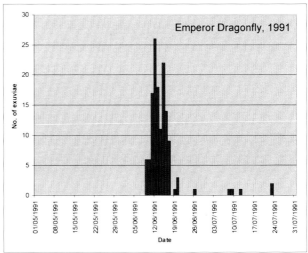

Figure 18. the emergence pattern of the Broad-bodied Chaser, *Libellula depressa* and the Emperor Dragonfly, *Anax imperator* for two consecutive years from the authors' garden pond.

the White-legged Damselfly, *Platycnemis pennipes* were collected during one season when only a few adults were seen.

The graphs in figure 18 show the emergence pattern of the Emperor Dragonfly, *Anax imperator* and the Broad-bodied Chaser, *Libellula depressa* for two consecutive years from the authors' garden pond.

At least eight species were known to visit the pond and seven species were observed ovipositing during the first year after the construction of the pond in 1989. Figure 19 shows the number of males to females of *Anax imperator* emerging and also indicates that greater numbers of males tend to emerge first followed by a higher ratio of females towards the end of the emergence period. The table in figure 20 shows the

total number of exuviae collected for 4 years following the construction of the pond. The increased numbers of *A.imperator* with the general decrease of the other species may suggest that *A.imperator* was preying on the other species. It was also noted that with the decrease of *A.imperator* in 1993 there was a considerable increase in the numbers of the Azure Damselfly, *Coenagrion puella*. With the maturing of the pond there was a decrease in the numbers of *L. depressa* which suggests that this species prefers new ponds or ponds where there is less aquatic vegetation and more exposed muddy areas at the bottom of the pond.

When collecting exuviae, care must be taken as they are easily damaged and blown away by the slightest breeze. The advantage of identifying exuviae

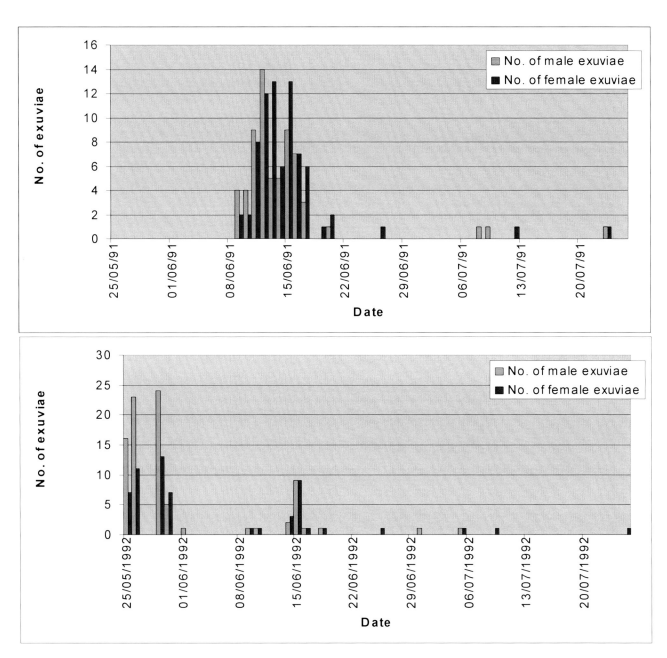

Figure 19. Total number of exuviae of *Anax imperator* during 1991 and 1992, showing the ratio of males to females on emergence.

	C. puella	L. depressa	I. elegans	A. imperator	S. striolatum	A. grandis	A. cyanea
1990	19	77	17	57	3		
1991	20	19	6	139	1		
1992	7	1	4	144			
1993	147	1	8	49	12	1	2

Figure 20. Total number of exuviae collected from a pond in the authors' garden over the 4-year period following its construction.

Searching for exuviae

instead of larvae is that they are easier to handle than larvae which, because they are alive, tend to wriggle and can make examination difficult. However, unlike larvae, exuviae have some disruption and damage to the dorsal surface of the thorax and the posterior area of the head and eyes due to the emergence of the adult dragonfly. Some exuviae may be covered in mud or debris that may obliterate key features for identification and in the case of damselfly exuviae, the caudal lamellae are often stuck together and twisted, or wholly missing. Lamellae that are stuck together or twisted can be gently teased apart using warm water and a fine artist's paint brush. Incomplete exuviae may make identification difficult or sometimes impossible.

Finding exuviae can be quite a challenge at first, but perseverance will soon help you to anticipate the species of dragonfly that should be present at a site, and where to search for their exuviae. Most exuviae are found on rushes and reeds in the marginal areas of water bodies, although the emerging larvae may use any vegetation or support. Experience will show that different species will choose specific niches within this general marginal area. However, this may vary from site to site and you should keep an open mind as to where exuviae may be found. Some exuviae of the Brilliant Emerald Dragonfly, *Somatochlora metallica,* the Southern Hawker, *Aeshna cyanea* and the Downy Emerald, *Cordulia aenea* are just three of the species which have been found on tree trunks or hanging from branches up in the trees. Exuviae of the Common Blue Damselfly, *Enallagma cyathigerum* have been found about 4 metres from water and the Black-tailed Skimmer, *Orthetrum cancellatum* up to 35 metres.

Exuviae key

The following simplified exuviae key is for the final instar larval skin from which the adult dragonfly has emerged, and is for those species that have been known to breed in Britain.

The anisopterans are generally easier to identify than the zygopterans because of their size and most can be identified in the field with the use of a x10 hand lens. Because of the variability of the length of the lateral spines of segments 8 and 9 of the Common Darter, *Sympetrum striolatum* and the Ruddy Darter, *S. sanguineum* determination may sometimes not be possible.

Lamellae are a key feature of identification for most of the zygopterans therefore, if this feature is missing, the use of a more detailed key and a microscope may be necessary. Examining the postocular region of the head, the occiput, is another feature that aids in the identification of the damselfly exuviae. Most are rounded and may be strongly spotted, indistinctly spotted or without spots (Miller, 1987), while a few are angular.

While it is possible to identify some zygopteran exuviae in the field, some may need to be scrutinised under a microscope with the use of a comprehensive key. The exuviae of the Small Red-eyed Damselfly, *Erythromma viridulum* look very similar to those of the Common Blue Damselfly, *Enallagma cyathigerum*. If the lamellae of either species are missing or if the characteristic narrow, dark transverse bands of the Common Blue Damselfly are indistinct or not evident, examination under a microscope is necessary to see the tiny stout setae on the ventral surface of the abdomen that are characteristic of the Small Red-eyed Damselfly.

The more experience you have examining exuviae with a key, the more proficient you will become. If at all in doubt about an identification use a more comprehensive guide to establish the identity by the use of additional features. Most keys are for the identification of larvae but can also be used for the final instar larval skin. There is a publication specifically for the identification of exuviae entitled *Die Exuvien Europaischer Libellen, The Exuviae of European Dragonflies* by Bernd Gerken & Klaus Sternberg. There is also a new photographic field guide to the anisopteran larvae and exuviae by Steve Cham. Both these books are listed in the references section.

Exuviae of Downy Emerald (N Donnithorne)

Simplified Exuviae Key

Developed by Gill Brook for use in the field. If, when using this key, there is any doubt as to the identification, a more detailed key should be used

Pictorial Glossary of Terms used in the Exuviae Key

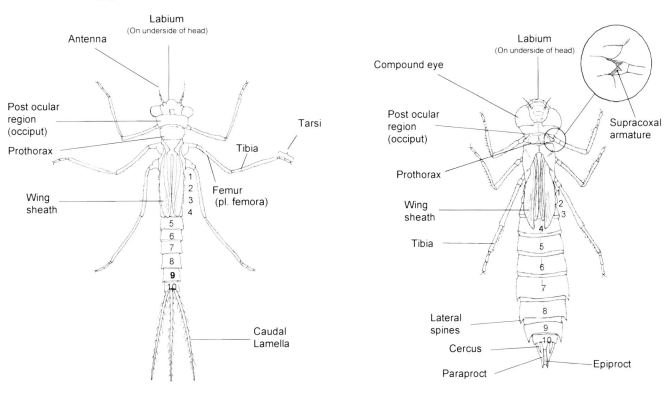

Zygoptera - Dorsal View

Labium (On underside of head)
Antenna
Post ocular region (occiput)
Prothorax
Wing sheath
Tarsi
Tibia
Femur (pl. femora)
Caudal Lamella

Anisoptera - Dorsal View

Labium (On underside of head)
Compound eye
Post ocular region (occiput)
Prothorax
Wing sheath
Tibia
Supracoxal armature
Lateral spines
Cercus
Paraproct
Epiproct

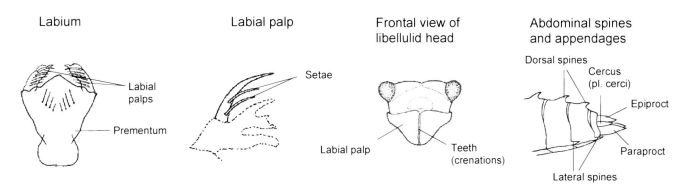

Labium

Labial palps
Prementum

Labial palp

Setae

Frontal view of libellulid head

Labial palp
Teeth (crenations)

Abdominal spines and appendages

Dorsal spines
Cercus (pl. cerci)
Epiproct
Paraproct
Lateral spines

EXUVIAE KEY

Zygoptera - slender and elongated body with 3 leaf-like lamellae (lamellae often stuck together and appearing as one)  page 100

Anisoptera - stout body ending in 5 spine-shaped appendages  page 102

Zygoptera

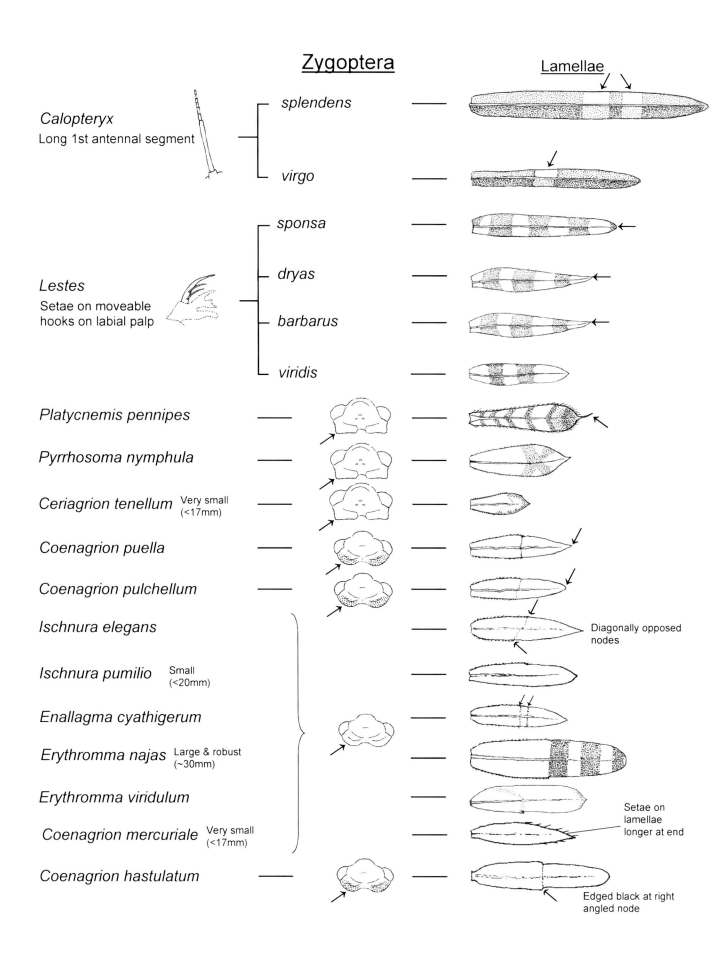

Lamellae

Calopteryx
Long 1st antennal segment

 splendens

 virgo

Lestes
Setae on moveable
hooks on labial palp

 sponsa

 dryas

 barbarus

 viridis

Platycnemis pennipes

Pyrrhosoma nymphula

Ceriagrion tenellum   Very small (<17mm)

Coenagrion puella

Coenagrion pulchellum

Ischnura elegans   Diagonally opposed nodes

Ischnura pumilio   Small (<20mm)

Enallagma cyathigerum

Erythromma najas   Large & robust (~30mm)

Erythromma viridulum   Setae on lamellae longer at end

Coenagrion mercuriale   Very small (<17mm)

Coenagrion hastulatum   Edged black at right angled node

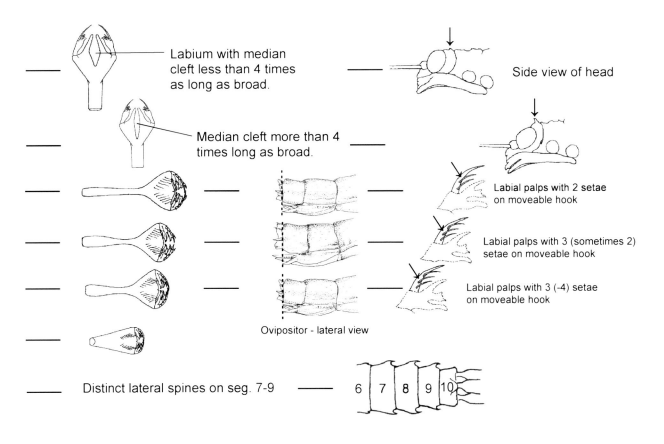

Labium with median cleft less than 4 times as long as broad.

Median cleft more than 4 times long as broad.

Side view of head

Labial palps with 2 setae on moveable hook

Labial palps with 3 (sometimes 2) setae on moveable hook

Labial palps with 3 (-4) setae on moveable hook

Ovipositor - lateral view

Distinct lateral spines on seg. 7-9

6 | 7 | 8 | 9 | 10

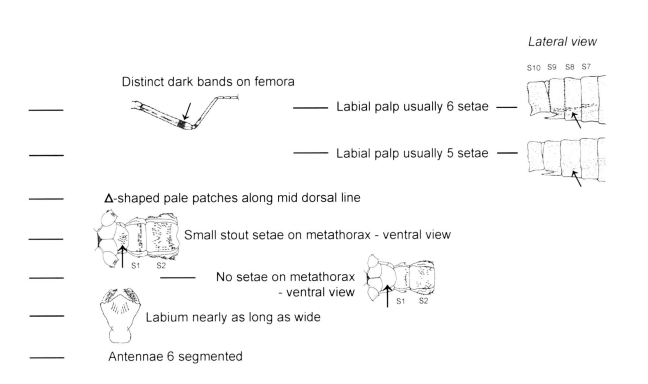

Lateral view

Distinct dark bands on femora

Labial palp usually 6 setae

S10 S9 S8 S7

Labial palp usually 5 setae

Δ-shaped pale patches along mid dorsal line

Small stout setae on metathorax - ventral view

S1 S2

No setae on metathorax - ventral view

S1 S2

Labium nearly as long as wide

Antennae 6 segmented

<u>Anisoptera</u>

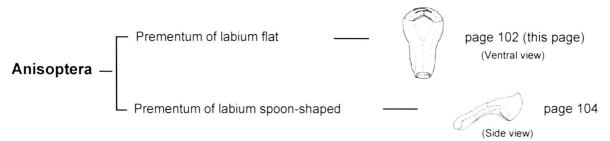

Anisoptera —

┌— Prementum of labium flat ———— page 102 (this page)
(Ventral view)

└— Prementum of labium spoon-shaped ——— page 104
(Side view)

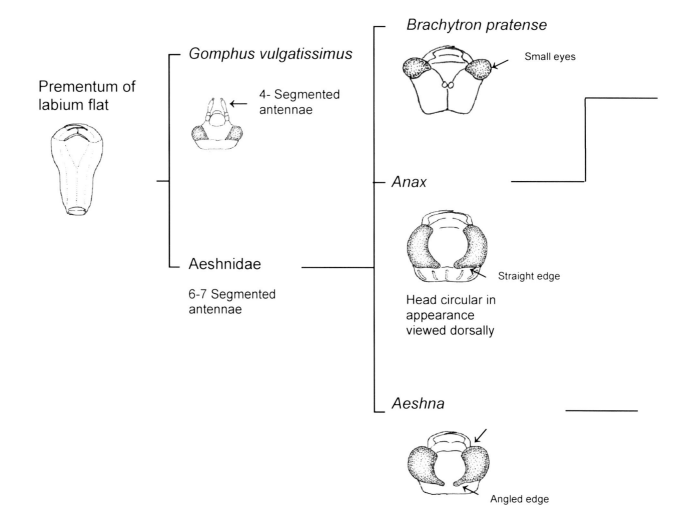

Prementum of labium flat

Gomphus vulgatissimus

4- Segmented antennae

Aeshnidae

6-7 Segmented antennae

Brachytron pratense

Small eyes

Anax

Straight edge

Head circular in appearance viewed dorsally

Aeshna

Angled edge

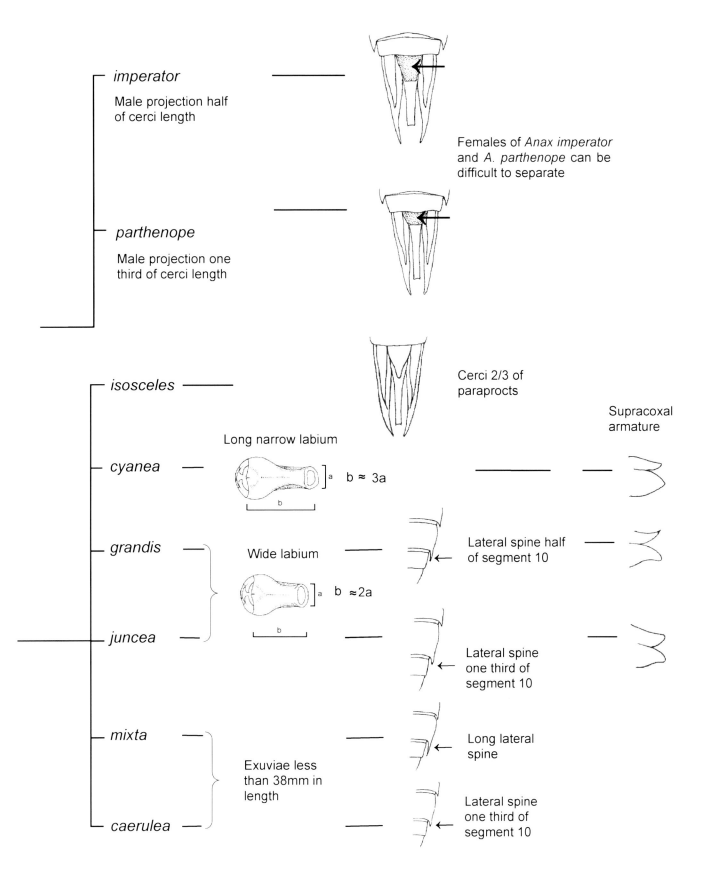

imperator

Male projection half
of cerci length

Females of *Anax imperator*
and *A. parthenope* can be
difficult to separate

parthenope

Male projection one
third of cerci length

isosceles

Cerci 2/3 of
paraprocts

Supracoxal
armature

Long narrow labium

cyanea

] a   b ≈ 3a

b

grandis

Wide labium

Lateral spine half
of segment 10

] a   b ≈ 2a

b

juncea

Lateral spine
one third of
segment 10

mixta

Long lateral
spine

Exuviae less
than 38mm in
length

caerulea

Lateral spine
one third of
segment 10

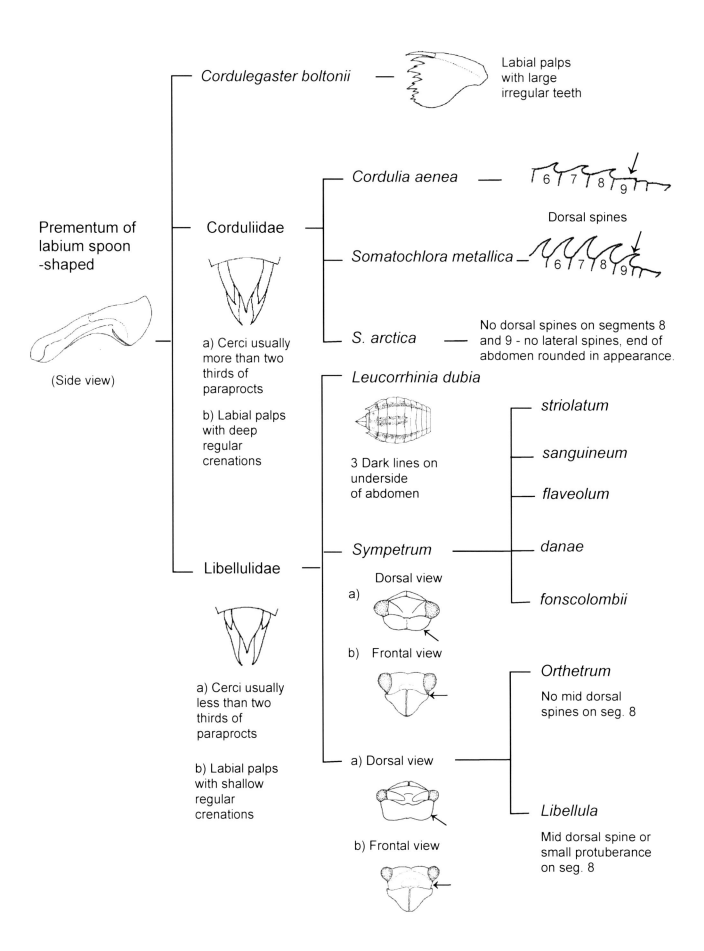

Prementum of labium spoon-shaped

(Side view)

Cordulegaster boltonii — Labial palps with large irregular teeth

Corduliidae

a) Cerci usually more than two thirds of paraprocts

b) Labial palps with deep regular crenations

Cordulia aenea — 6 7 8 9

Dorsal spines

Somatochlora metallica — 6 7 8 9

S. arctica — No dorsal spines on segments 8 and 9 - no lateral spines, end of abdomen rounded in appearance.

Libellulidae

a) Cerci usually less than two thirds of paraprocts

b) Labial palps with shallow regular crenations

Leucorrhinia dubia

3 Dark lines on underside of abdomen

Sympetrum

a) Dorsal view

b) Frontal view

striolatum

sanguineum

flaveolum

danae

fonscolombii

a) Dorsal view

b) Frontal view

Orthetrum

No mid dorsal spines on seg. 8

Libellula

Mid dorsal spine or small protuberance on seg. 8

———— Dark markings on side of thorax

———— Fuzzy tufts on occiput

a = seg. 9, b = spine
c = seg. 8, d = spine

———— Mid dorsal spines on seg. 5-8 ———— b>a ———— d = more than 1/3 of c viewed ventrally

———— Mid dorsal spines on seg. 5-8 ———— b<a ———— d = 1/3 or less of c viewed ventrally

———— Mid dorsal spines on seg. 5 or 6-8 ———— b ≈ ½ a ———— b = almost twice d

———— Mid dorsal spines on seg. 5-7 ———— <u>Very</u> short lateral spines

———— No dorsal spines

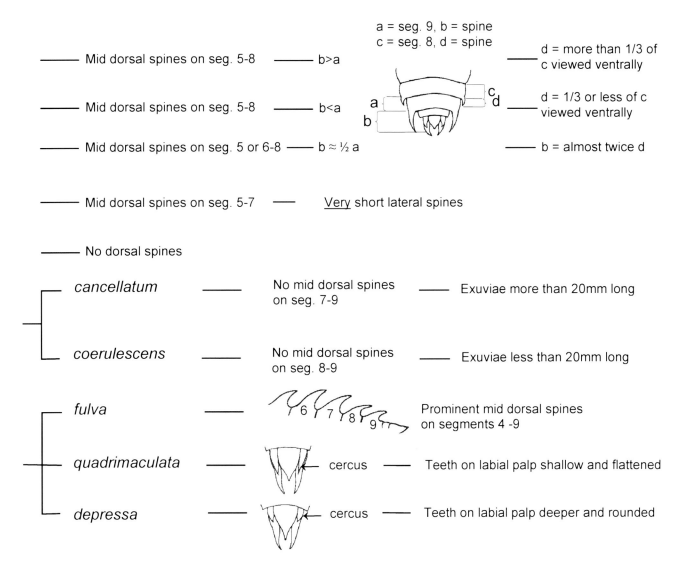

cancellatum ———— No mid dorsal spines on seg. 7-9 ———— Exuviae more than 20mm long

coerulescens ———— No mid dorsal spines on seg. 8-9 ———— Exuviae less than 20mm long

fulva ———— Prominent mid dorsal spines on segments 4 -9

quadrimaculata ———— cercus ———— Teeth on labial palp shallow and flattened

depressa ———— cercus ———— Teeth on labial palp deeper and rounded

A selection of sites to visit

The following sites have been chosen because between them they support all the 27 resident Kent species and have public access. It may also be possible, given suitable weather conditions and good information, to see some of the seven migrant species.

A brief description of each site is followed by a list of the species recorded there.

* - Exuviae found

\# - Migrant species

1. Dungeness. Gravel pit lakes and ponds.

English Nature and Royal Society for the Protection of Birds. Parking at TR068185 (free) (RSPB fee for admission to reserve, free for members), TR088169 (free) and TR063197 (free). 22 species.

The Dungeness area is one of Europe's largest shingle expanses and one of the most important shingle beach habitats in Europe supporting a large number of specialised species of flora and fauna. Much of this SSSI is open to public access. To visit the RSPB Reserve drive to the RSPB car park which can be reached by turning south at Boulderwall Farm. There are several public footpaths from Boulderwall Farm giving access to good pond sites. There is now also a car park on the north side of the Lydd to Dungeness road opposite Boulderwall Farm. From this car park, there is access to more gravel pit lakes and pools where *Anax parthenope* and *Sympetrum fonscolombii* are sometimes seen. To the east of the RSPB reserve, at TR0818, are the Long Pits that are well worth a visit. They have well vegetated margins, and are surrounded by bushes and scrub which provide sheltered areas for dragonflies. These long and narrow water bodies can be reached from the road from Lydd to Dungeness, or from the old lighthouse car park. Because of the geographical position Dungeness is a good place to see migrants especially when there are prolonged warm easterly winds, as in

1995. This site, Sandwich Bay and Scotney Castle have the greatest diversity of dragonflies in Kent.

Site list

Emerald Damselfly *Lestes sponsa* *
Red-eyed Damselfly *Erythromma najas* *
Small Red-eyed Damselfly *E. viridulum*
Azure Damselfly *Coenagrion puella*
Variable Damselfly *C. pulchellum*
Common Blue Damselfly *Enallagma cyathigerum* *
Blue-tailed Damselfly *Ischnura elegans* *
Migrant Hawker *Aeshna mixta* *
Southern Hawker *A. cyanea*,
Brown Hawker *A. grandis* *
Emperor Dragonfly *Anax imperator* *
Lesser Emperor *A. parthenope* #
Hairy Dragonfly *Brachytron pratense*
Four-spotted Chaser *Libellula quadrimaculata*,
Broad-bodied Chaser *L. depressa*,
Black-tailed Skimmer *Orthetrum cancellatum* *
Common Darter *Sympetrum striolatum* *
Vagrant Darter *S. vulgatum* #
Red-veined Darter   *S. fonscolombii* *#
Yellow-winged Darter *S. flaveolum* #
Ruddy Darter *S. sanguineum* *
Black Darter *S. danae* #

Former gravel pit. Dungeness

Scotney Castle

2. Scotney Castle, moat, streams and nearby river.

National Trust, parking at N.T. car park TQ686353 (free) (admission charge to property). 22 species.

This is a National Trust property which has a moat round the Old Castle with nearby streams. The moat has abundant marginal vegetation and surface plants such as water lilies. Hornwort forms a mat at the surface suitable for *Erythromma viridulum*. A small stream that runs through the southern end of the garden and enters the moat at its south-west corner is often frequented by *Calopteryx splendens* and *C. virgo*. The River Bewl runs along the southeast border of the grounds and is joined by a small stream, the Sweet Bourne, running along the south-western side. Near this junction the Sweet Bourne is deep cut and looks a suitable breeding habitat for *Cordulegaster boltonii*. A small number of *Somatochlora metallica* breed at Collier's Pond that is in the woodland on the north side of the long driveway to the NT property. Much of this large pond is inaccessible due to dense undergrowth and steep sides. This site and Dungeness have the greatest diversity of resident dragonfly species in Kent.

Site list
Beautiful Demoiselle *Calopteryx virgo*,
Banded Demoiselle *C. splendens*,
White-legged Damselfly *Platycnemis pennipes* *
Large Red Damselfly *Pyrrhosoma nymphula*
Red-eyed Damselfly *Erythromma najas* *
Small Red-eyed Damselfly *E. viridulum*,
Azure Damselfly *Coenagrion puella* *
Common Blue Damselfly *Enallagma cyathigerum* *
Blue-tailed Damselfly *Ischnura elegans* *
Migrant Hawker *Aeshna mixta*
Southern Hawker *A. cyanea* *
Brown Hawker *A. grandis*
Emperor Dragonfly *Anax imperator*
Hairy Dragonfly *Brachytron pratense*,
Golden-ringed Dragonfly *Cordulegaster boltonii*
Downy Emerald *Cordulia aenea* *
Brilliant Emerald *Somatochlora metallica*,*
Four-spotted Chaser *Libellula quadrimaculata*
Broad-bodied Chaser *L. depressa*
Black-tailed Skimmer *Orthetrum cancellatum*
Common Darter *Sympetrum striolatum* *
Ruddy Darter *S. sanguineum* *

Delf Stream, Sandwich

3. Sandwich Bay. Ponds, river, dykes and nature reserve.

Sandwich Bay Bird Observatory Trust. Parking at the Bird Observatory (TR355575) free, but a reduced road toll for those visiting the Bird Observatory. 22 species.

Sandwich Bay is a large coastal site comprised mainly of sand dunes, grassland and occasional scrub. It is a good site for orchids and birds as well as dragonflies. There are various waterbodies including Restharrow Scrape, Middle Field Pond, North Stream, which has moderate flow, and Delf Stream. Because of the nearness to the continent migrants can sometimes be seen at Sandwich Bay. Restharrow Scrape is a large man-made area of shallow water where *Sympetrum fonscolombei* is probably breeding. This area and the new Middle Field Pond have provided suitable habitats for a greater diversity of species. In the past *Libellula fulva* has been recorded at North Stream. As Sandwich Bay is a private estate it is necessary to stay on the public footpaths, except in the area of the nature reserve. Photocopied maps of the area may be obtained from the visitor centre.

Site list
Emerald Damselfly *Lestes sponsa*
Large Red Damselfly *Pyrrhosoma nymphula*
Red-eyed Damselfly *Erythromma najas*
Small Red-eyed Damselfly *E. viridulum*
Azure Damselfly *Coenagrion puella*
Variable Damselfly *C. pulchellum*
Common Blue Damselfly *Enallagma cyathigerum* *
Blue-tailed Damselfly *Ischnura elegans*
Migrant Hawker *Aeshna mixta*
Southern Hawker *A. cyanea*
Brown Hawker *A. grandis*
Emperor Dragonfly *Anax imperator* *
Hairy Dragonfly *Brachytron pratense*
Four-spotted Chaser *Libellula quadrimaculata*
Scarce Chaser *L. fulva*
Broad-bodied Chaser *L. depressa*
Black-tailed Skimmer *Orthetrum cancellatum*
Common Darter *Sympetrum striolatum* *
Red-veined Darter *S. fonscolombei* #
Yellow-winged Darter *S. flaveolum* #
Ruddy Darter *S. sanguineum*
Black Darter *S. danae* #

Marshall's Lake at Bedgebury Pinetum

4. Bedgebury. Pinetum lakes, ponds, and stream; Forest lake and stream.

Forestry Commission. Parking at TQ716332 (car park fee), or TQ741335 (free). 21 species.

The lakes, ponds and stream in the Pinetum, and Louisa Lake and outflow stream in the forest, can be visited together on foot as they are only a pleasant one and a half miles forest walk apart. Alternatively, they can be visited separately from one or the other of the parking places. In the Pinetum, the national conifer collection, there is the ornamental Marshall's Lake, with large drifts of water lilies where *Erythromma najas* has been known to breed. Exuviae of *Cordulia aenea*, *Somatochlora metallica*, and *Platycnemis pennipes* have also been found there. An additional lake has been constructed by the new visitor centre. Between these two lakes there are two large ponds, one of which has a variable level and a greater diversity of species. A small, almost invisible stream connects the two ponds, and it is here where exuviae of *Cordulegaster boltonii* have been collected. This species also breeds in the tiny outflow stream from Louisa Lake, in the forest, which is a mature naturalised lake formed by the damming of a tiny stream with a small concrete weir. Bog Pondweed *Potamogeton polygonifolius* Pourret is abundant at the edges showing the water to be acidic.

Site list
Beautiful Demoiselle *Calopteryx virgo*
Banded Demoiselle *C. splendens*
Emerald Damselfly *Lestes sponsa* *
White-legged Damselfly *Platycnemis pennipes* *
Large Red Damselfly *Pyrrhosoma nymphula* *
Red-eyed Damselfly *Erythromma najas* *
Azure Damselfly *Coenagrion puella* *
Common Blue Damselfly *Enallagma cyathigerum* *
Blue-tailed Damselfly *Ischnura elegans* *
Migrant Hawker *Aeshna mixta* *
Southern Hawker *A. cyanea* *
Brown Hawker *A. grandis* *
Emperor Dragonfly *Anax imperator* *
Golden-ringed Dragonfly *Cordulegaster boltonii* *
Downy Emerald *Cordulia aenea* *
Brilliant Emerald *Somatochlora metallica* *
Four-spotted Chaser *Libellula quadrimaculata* *
Broad-bodied Chaser *L. depressa* *
Black-tailed Skimmer *Orthetrum cancellatum* *
Common Darter *Sympetrum striolatum* *
Ruddy Darter *S. sanguineum* *

Pond at Sevenoaks Wildlife Reserve

5. Sevenoaks Wildlife Reserve. River, lakes and ponds.

Kent Wildlife Trust. Parking at TQ519566. 20 species.

Developed from a former large quarry site, this reserve, which is popular with birdwatchers and anglers, has bird hides overlooking some of the lakes. *Cordulia aenea* has been observed at one of the smaller lakes. The fast clear waters of the River Darent pass through the site and there are also some small ponds adding to the variety of habitats suitable for a good diversity of dragonfly species and many other wildlife groups.

Site list

Beautiful Demoiselle *Calopteryx virgo*
Banded Demoiselle *C. splendens* *
Emerald Damselfly *Lestes sponsa* *
White-legged Damselfly *Platycnemis pennipes*
Large Red Damselfly *Pyrrhosoma nymphula* *
Red-eyed Damselfly *Erythromma najas* *
Small Red-eyed *Damselfly E. viridulum*
Azure Damselfly *Coenagrion puella* *
Common Blue Damselfly *Enallagma cyathigerum* *
Blue-tailed Damselfly *Ischnura elegans* *
Migrant Hawker *Aeshna mixta* *
Southern Hawker *A. cyanea* *
Brown Hawker *A. grandis* *
Emperor Dragonfly *Anax imperator* *
Downy Emerald *Cordulia aenea* *
Four-spotted Chaser *Libellula quadrimaculata* *
Broad-bodied Chaser *L. depressa*
Black-tailed Skimmer *Orthetrum cancellatum* *
Common Darter *Sympetrum striolatum* *
Ruddy Darter *S. sanguineum* *

Fishing Lake at Shorne Wood Country Park

6. Shorne Wood Country Park. Ponds and ditches.

Kent County Council & Gravesham Borough Council. Parking at TQ684698 (fee). 20 species.

This 70 ha (174 acres) site is a mixture of broadleaved woodland, large open grassy areas and old quarry earth-works clothed in silver birch scrub. There are two fishing lakes, where *Erythromma viridulum* may be seen, several ponds, ditches and a shallow flooded boggy area. It is a busy and popular place for walking and recreation, especially at weekends and in the summer. There is also an information centre.

Site list

Banded Demoiselle *Calopteryx splendens* *
Emerald Damselfly *Lestes sponsa* *
Scarce Emerald Damselfly *L. dryas*
Large Red Damselfly *Pyrrhosoma nymphula* *
Red-eyed Damselfly *Erythromma najas* *
Small Red-eyed Damselfly *E. viridulum* *
Azure Damselfly *Coenagrion puella* *
Common Blue Damselfly *Enallagma cyathigerum* *
Blue-tailed Damselfly *Ischnura elegans* *
Migrant Hawker *Aeshna mixta* *
Southern Hawker *A. cyanea* *
Brown Hawker *A. grandis* *
Emperor Dragonfly *Anax imperator* *
Hairy Dragonfly *Brachytron pratense* *
Downy Emerald *Cordulia aenea*
Four-spotted Chaser *Libellula quadrimaculata* *
Broad-bodied Chaser *L. depressa*
Black-tailed Skimmer *Orthetrum cancellatum* *
Common Darter *Sympetrum striolatum* *
Ruddy Darter *S. sanguineum* *

Stream at Sissinghurst Castle which supports Golden-ringed Dragonfly and Beautiful Demoiselle

7. Sissinghurst Castle. Lakes and stream.

National Trust. Parking at N.T. car park TQ806384 (fee for non NT members). 19 species.

To the southeast of the famous Sissinghurst Castle Gardens lie the two lakes, with a meadow and field on their northern sides, and broadleaved woodland to the south. The margins are well vegetated with rushes, reeds and sedges, with some trees at the water's edge. The eastern lake has one island and the western lake has two. At the western end of the latter, is an area of Great Reedmace *Typha latifolia* (Linnaeus) and marsh. *Cordulia aenea* and *Platycnemis pennipes* breed at both of these lakes. A small narrow stream runs along the southern side of the lakes from which it is partly fed. *Calopteryx virgo* occurs here and may be breeding in this stream. *Cordulegaster boltonii* has also been seen flying along the stream which would appear to be a suitable breeding habitat, but although no exuviae have been found here, they have been found about half a mile upstream. The National Trust

at Sissinghurst would welcome records of any new species seen at this site.

Site list
Beautiful Demoiselle *Calopteryx virgo*
Banded Demoiselle *C.splendens*
White-legged Damselfly *Platycnemis pennipes* *
Large Red Damselfly *Pyrrhosoma nymphula* *
Red-eyed Damselfly *Erythromma najas* *
Azure Damselfly *Coenagrion puella* *
Common Blue Damselfly *Enallagma cyathigerum* *
Blue-tailed Damselfly *Ischnura elegans* *
Migrant Hawker *Aeshna mixta* *
Southern Hawker *A. cyanea* *
Brown Hawker *A. grandis*
Emperor Dragonfly *Anax imperator* *
Hairy Dragonfly *Brachytron pratense* *
Golden-ringed Dragonfly *Cordulegaster boltonii* *
Downy Emerald *Cordulia aenea* *
Four-spotted Chaser *Libellula quadrimaculata*
Black-tailed Skimmer *Orthetrum cancellatum* *
Common Darter *Sympetrum striolatum* *
Ruddy Darter *S. sanguineum*

8. Hothfield Common. Bogs and ponds.

Local Nature Reserve, Kent Wildlife Trust and Ashford Borough Council. Parking at TQ972458. 17 species.

Hothfield Common is the largest piece of heathland in Kent and it is here where the best example of a true bog in the county may be found. At the west end of the main bog there is a small pond. The main bog has many typical acid loving plants such as Bog Asphodel *Narthecium ossifragum* (Linnaeus) Hudson, Round-leaved Sundew *Drosera rotundifolia* Linnaeus, *Sphagnum* spp. and Bog Pondweed *Potamogeton polygonifolius* Pourret. In recent years the Kent Wildlife Trust has improved the main bog by felling many trees that threatened its existence. Cattle now graze the heathland, which helps to prevent tree seedlings from becoming established. This very important place is the only site in the county where *Orthetrum coerulescens* breeds. A raised walk crosses the main bog allowing a closer and better view of this dragonfly.

Site list

Banded Demoiselle *Calopteryx splendens*
Emerald Damselfly *Lestes sponsa*
Large Red Damselfly *Pyrrhosoma nymphula* *
Azure Damselfly *Coenagrion puella*
Common Blue Damselfly *Enallagma cyathigerum* *
Blue-tailed Damselfly *Ischnura elegans* *
Migrant Hawker *Aeshna mixta*
Southern Hawker *A. cyanea* *
Brown Hawker *A. grandis*
Emperor Dragonfly *Anax imperator* *
Hairy Dragonfly *Brachytron pratense*
Four-spotted Chaser *Libellula quadrimaculata* *
Broad-bodied Chaser *L. depressa* *
Keeled Skimmer *Orthetrum coerulescens* *
Common Darter *Sympetrum striolatum* *
Yellow-winged Darter *S. flaveolum* #
Ruddy Darter *S. sanguineum* *

River Great Stour at Westbere

9. Westbere. River, gravel pit lakes and dykes.

Limited parking at TR196610. 17 species.

This site is comprised of three habitats; the Great Stour river, flooded gravel pits and densely vegetated marsh with a few dykes. The whole area has many trees providing sunny sheltered places.

From the parking place cross the railway line and follow the track, that goes southeast towards the river. On both sides of this track are well-vegetated, fresh water dykes reminiscent of fenland. If *Coenagrion pulchellum* is on the wing, it may be seen here. At the end of the track, you reach the River Great Stour with its clear water of moderate flow. To the east are the marshes and river and to the west is the better riverside path. The banks are steep in places and well vegetated with reeds, rushes, bushes and nettles. It was at an accessible point that the authors found exuviae of *Libellula fulva* on large sedges (*Carex* spp). This scarce dragonfly may be seen at any of these locations from the railway crossing onwards. On the west side of the first track, between the railway line and the river, is a huge flooded gravel pit with several private access points which, unfortunately, are only open to anglers with permits. This lake is a good site for dragonflies that may be seen as they fly and settle along the riverside path.

Site list
Banded Demoiselle *Calopteryx splendens* *
Emerald Damselfly *Lestes sponsa*
Large Red Damselfly *Pyrrhosoma nymphula*
Red-eyed Damselfly *Erythromma najas* *
Azure Damselfly *Coenagrion puella* *
Variable Damselfly *C. pulchellum* *
Common Blue Damselfly *Enallagma cyathigerum* *
Blue-tailed Damselfly *Ischnura elegans* *
Migrant Hawker *Aeshna mixta* *
Southern Hawker *A. cyanea*
Brown Hawker *A. grandis*
Emperor Dragonfly *Anax imperator* *
Hairy Dragonfly *Brachytron pratense* *
Scarce Chaser *Libellula fulva* *
Black-tailed Skimmer *Orthetrum cancellatum* *
Common Darter *Sympetrum striolatum* *
Ruddy Darter *Sympetrum sanguineum*

10. Haysden Country Park. River, lakes and streams.

Parking at TQ571459.   Tonbridge and Malling Council. 17 species.

This country park is popular for walking and recreation, including fishing and water sports. There are 65ha (160 acres) of parkland, woodland and lakes, and the River Medway runs through the length of the park. A few years ago, part of the river course was diverted and a flood barrier built to protect Tonbridge. Many damselflies, including *Calopteryx virgo*, may be seen by the old river course, which is near the car park.

Site list

Beautiful Demoiselle *Calopteryx virgo*
Banded Demoiselle *C. splendens*
White-legged Damselfly *Platycnemis pennipes* *
Large Red Damselfly *Pyrrhosoma nymphula* *
Red-eyed Damselfly *Erythromma najas* *
Small Red-eyed Damselfly *E. viridulum*
Azure Damselfly *Coenagrion puella*
Common Blue Damselfly *Enallagma cyathigerum* *
Blue-tailed Damselfly *Ischnura elegans* *
Migrant Hawker *Aeshna mixta* *
Southern Hawker *A. cyanea*
Brown Hawker *A. grandis*
Emperor Dragonfly *Anax imperator*
Broad-bodied Chaser *Libellula depressa*
Black-tailed Skimmer *Orthetrum cancellatum*
Common Darter *Sympetrum striolatum*
Ruddy Darter *S.sanguineum*

11. The Royal Military Canal.

The Romney Marsh Countryside Project and National Trust. 17 species.

The Kentish section of this canal, which was built because of the threat of a Napoleonic invasion, is approximately twenty miles long. The RMC Path follows the entire length of the canal from Seabrook in Kent to its end in East Sussex. Observations can also be made from any of the dozen or so bridges that cross over this waterway. There is a three and a half mile section east from Appledore that includes a dyke parallel to the RMC. The canal is a relatively static, linear freshwater habitat. Fringed Waterlily *Nymphoides peltata* Kuntze is quite prolific in places and makes a suitable breeding site for *Erythromma najas*. It is also a good place to see *Brachytron pratense* early in the season, and exuviae can often be found on the banks clinging to marginal vegetation.

Site list

Banded Demoiselle *Calopteryx splendens*
Emerald Damselfly *Lestes sponsa* *
Large Red Damselfly *Pyrrhosoma nymphula*
Red-eyed Damselfly *Erythromma najas* *
Small Red-eyed Damselfly *E. viridulum*
Azure Damselfly *Coenagrion puella* *
Common Blue Damselfly *Enallagma cyathigerum* *
Blue-tailed Damselfly *Ischnura elegans* *
Migrant Hawker *Aeshna mixta*
Southern Hawker *A.cyanea*
Brown Hawker *A. grandis*
Emperor Dragonfly *Anax imperator*
Hairy Dragonfly *Brachytron pratense* *
Four-spotted Chaser *Libellula quadrimaculata*
Black-tailed Skimmer *Orthetrum cancellatum*
Common Darter *Sympetrum striolatum* *
Ruddy Darter *S. sanguineum*

12. Cliffe Pools and Marshes. Gravel pit lakes and dykes.

RSPB. Parking at TQ736766 near church or along rough track from TQ736768 to TQ714777. 17 species.

This Thameside site consists of large gravel pit lakes, marsh grasslands and dykes. The track to the site of the now demolished Coastguard Cottages (TQ714776) is rough but worth the effort for a chance to see *Lestes dryas*. This rare damselfly may be present all along the northern dykes of this area from Cliffe eastwards to Grain. The authors have found this species near the estuary end of the rough track in the well-vegetated, brackish dykes. There is some concern for the survival of this species as in recent hot dry summers these dykes tend to dry up. Exuviae of *L.viridis* were found in this location in 1992. There are also footpaths and tracks by some of the lakes, which are a nature reserve and popular to bird watchers.

Site list

Willow Emerald Damselfly *Lestes viridis* *#
Emerald Damselfly *Lestes sponsa* *
Scarce Emerald Damselfly *Lestes dryas* *
Azure Damselfly *Coenagrion puella* *
Common Blue Damselfly *Enallagma cyathigerum*
Blue-tailed Damselfly *Ischnura elegans* *
Migrant Hawker *Aeshna mixta* *
Southern Hawker *A. cyanea*
Brown Hawker *A. grandis*
Emperor Dragonfly *Anax imperator* *
Four-spotted Chaser *Libellula quadrimaculata* *
Black-tailed Skimmer *Orthetrum cancellatum*
Common Darter *Sympetrum striolatum* *
Red-veined Darter *S. fonscolombii* #
Yellow-winged Darter *S. flaveolum* #
Ruddy Darter *S. sanguineum* *
Black Darter *S. danae* #

Notable events in recent years

2006 migrant influx

From among the records received during 2006, there are a few worthy of mention. The hot dry summer of 2006, with easterly winds, saw an influx of migrants from the nearby continent into Britain. Kent had a small share of these.

The Lesser Emperor, *Anax parthenope*, continued to be recorded at the RSPB Reserve at Dungeness. Unfortunately, no exuviae of this species have yet been found there and so it is still uncertain as to whether these individuals were local progeny or migrants. An immature male *A. parthenope* was seen and photographed at Dartford on 23 June.

Up to six Red-veined Darters, *Sympetrum fonscolombii*, were also recorded from the Dungeness RSPB Reserve on the 12-13 June. Although breeding has been recorded at this site, possibly since 1995 (see species account), no exuviae of this species were reported in 2006 so it is unclear as to whether these individuals were locally bred or were migrants. In mid June 15 males *S. fonscolombii* were seen at Restharrow Scrape, Sandwich Bay and on 25 August 6-8 individuals were also seen at this site. A further sighting for this species was at Bank Sand Point by the River Great Stour on 14 July.

The Yellow-winged Darter, *S. flaveolum*, was seen in July at three sites by the coast. One individual was seen at Herne Bay Downs on 19 July, another at North Foreland on 23 July, and 17 at Kingsgate Golf Course on 21 July. The latter were seen along with one Black Darter, *S. danae*, as well as with the Common Darter, *S. striolatum*, and the Ruddy Darter, *S. Sanguineum*.

2007/2008

2007 was a remarkable year for early emergents due to the hottest April on record with temperatures most days being above 20°C. In parts of Britain some dragonflies were seen as early as March. In Kent the first known record was of the Large Red Damselfly, *Pyrrhosoma nymphula*, on 12 April. Other notable early records were the emergence of the Emperor Dragonfly, *Anax imperator*, on 24 April and *Cordulia aenea* on 26 April.

In 2008 the authors were contacted by Ian Hodgson, warden of the Sandwich Bay Bird Observatory, informing us that numerous *Libellula fulva* had been seen again at various locations at Sandwich Bay, including a female ovipositing at North Stream. It was good to hear of the return of this dragonfly to Sandwich Bay after many years of absence. On 12 July, the authors visited the site on a cloudy day but unfortunately no *L. fulva* were seen, although they were rewarded by a resting *Anax parthenope* in the nature reserve. However on a second visit on 22 July with Ian many male *L. fulva* were seen at North Stream and Delf Stream. Another important record was of a female *Lestes dryas* (Scarce Emerald Damselfly) photographed on 30 July 2008 on private land near Horsmonden. The only other records of this nationally scarce species in Kent are from the north of the county.

2009 Update

The Lesser Emperor, *Anax parthenope*, was seen again at Sandwich Bay Bird Observatory for the second year running. It may have arrived from the continent on easterly winds, along with Red-veined Darters, *Sympetrum fonscolombii*, which occurred in increased numbers at the Observatory. The Variable Damselfly, *Coenagrion pulchellum*, continues to spread to new sites on the nature reserve.

The Scarce Chaser, *Libellula fulva*, was seen at the Sandwich reserve again this year. This species has been spreading in parts of England. In Kent, two new sightings were reported in West Kent near the River Eden. Another species which is spreading in the county is the Downy Emerald, *Cordulia aenea*, with sites in two new 10km squares.

References

Askew, R. R. (1988) *The Dragonflies of Europe.* Harley Books, Colchester.

Attridge, W. (1995) The August Influx 1995. *Dungeness Bird Observatory Bulletin* 40: 8-9.

Attridge, W. (1996) The Dragonflies of the Dungeness Bird Observatory Recording Area - 1995. Dungeness Bird Observatory Report for 1995: 80-82.

Attridge, W. (2000) The Dragonflies of the Dungeness Bird Observatory Recording Area - 1999. Dungeness Bird Observatory Report for 1999: 75-76.

Bath, W. H. (1890) *An Illustrated Handbook of British Dragonflies.* Birmingham.

Brook, G. (2003) Identification of the exuvia of the Small Red-eyed Damselfly *Erythromma viridulum* (Charpentier). *Journal of the British Dragonfly Society* 19: 40-43.

Brook, G. & Brook, J. (1996) Odonata Report 1995. *The Bulletin of the Kent Field Club* No. 41: 53.

Brook, G. & Brook, J. (2003) The Willow Emerald Damselfly *Chalcolestes viridis* (Vander Linden) in Kent. *Journal of the British Dragonfly Society* 19: 51-54.

Brook, G. & Brook, J. (2004) Past Breeding Evidence of Willow Emerald Damselfly *Chalcolestes viridis* (Vander Linden) in Kent. *Atropos* 21: 3-5.

Brooks, S. (1997) *Field Guide to the Dragonflies and Damselflies of Great Britain and Ireland.* British Wildlife Publishing, Hook, Hampshire.

Brownett, A. (1996) *The Dragonflies of Oxfordshire.* Brookside Books, Banbury.

Cham, S. (2002). The range expansion of Small Red-eyed Damselfly *Erythromma viridulum* (Charp.) in the British Isles. *Atropos* 15: 3-9.

Cham, S. (2007) *Field Guide to the larvae and exuviae of British Dragonflies. Volume 1: Dragonflies (Anisoptera).* The British Dragonfly Society. Gem Publishing Company, Brightwell, Wallingford.

Chelmick, D., Hammond, C., Moore, N. & Stubbs, A. (1980) *The Conservation of Dragonflies.* Nature Conservancy Council, London.

Corbet, P. S. (1957) The Life-history of the Emperor Dragonfly *Anax imperator* Leach (Odonata : Aeshnidae). *Journal of Animal Ecology* 26:1-69.

Corbet P. S., Longfield, C. & Moore, N. W. (1960) *Dragonflies.* Collins, London.

Corbet, P. S. (1999) *Dragonflies; behaviour and ecology of Odonata.* Harley Books, Colchester.

Crowson, R. A. (1946) The fossil insects of the Weald. in Given, J. C. M. (Ed) *Royal Tunbridge Wells - Past and Present.* Tunbridge Wells, Courier.

d'Aguilar, J. , Dommanget, J-L. & Prechac, R. (1986) English Edition. *A Field Guide to the Dragonflies of Britain, Europe and North Africa.* Collins, London.

Dannreuther, T. (1939) The Dragonflies of East Sussex. *The Hastings and East Sussex Naturalist,* Vol. VI. (6): 272-284.

Dannreuther, T. (1941) p. 64 in *South-east Naturalist.*

Dijkstra, K-D. B. (2006) *Field Guide to the Dragonflies of Britain and Europe.* British Wildlife Publishing, Gillingham, Dorset.

Evans, W. F. (1845) *British Libellulinae or Dragonflies.* London.

Follett, P. (1996) *Dragonflies of Surrey.* Surrey Wildlife Trust, Woking.

Gerken, B. & Sternberg, K., (1999). *Die Exuvien Europaischer Libellen.* Huxaria Druckerei Gmb H.

Hammond, C. O. (1983) *The Dragonflies of Great Britain and Ireland.* Revised Edition. Harley Books, Colchester.

Longfield, C. (1949a) *The Dragonflies of the British Isles.* Second Edition. Warne, London.

Longfield, C. (1949b) The dragonflies of the London area. *London Naturalist* 28:80-98.

Lucas, W. J. (1900) *British Dragonflies.* Upcott Gill, London.

McGeeney, A. (1997) *Aeshna mixta*, pp. 104-105 in Brooks, S. (Ed.) *Field Guide to the Dragonflies and Damselflies of Great Britain and Ireland*. British Wildlife Publishing, Hook.

McLachlan, R. (1884) The British dragon-flies annotated. *Entomologists Monthly Magazine* 20:251-256.

Merritt, R., Moore, N.W. & Eversham, B.C. (1996) *Atlas of the dragonflies of Britain and Ireland*. HMSO, London.

Miller, P. L. (1995) *Dragonflies*. Naturalists' Handbooks 7. Richmond Publishing Co. Ltd. Slough.

Page,W. (1908) *The Victoria History of the County of Kent*. Vol. 1. Constable, London.

Parr, A. (1999) Migrant Dragonflies in 1998. *Atropos* 6:70.

Parr, A. (2001) Re: *Pantala flavescens* in Kent (e-mail to J.S. Badmin, dated 10.05.01)

Prendergast, E. D. V. (1991) *The Dragonflies of Dorset*. Dorset Natural History and Archaeological Society, Dorchester.

Stephens, J. F. (1835-1837) *Illustrations of British Entomology*. Baldwin & Cradock, London.

Ullyett, H. (1888) *Rambles of a Naturalist Round Folkestone*. J. English, Folkestone.

Vick, G. (1997) *Somatochlora metallica*, pp. 126-127 in Brooks, S. (Ed.) *Field Guide to the Dragonflies and Damselflies of Great Britain and Ireland*. BritishWildlife Publishing, Hook.

Uteeri, C., Carchini, G., Falchelti, E. & Belfiore, C. (1984) Philopatry, homing and dispersal in *Lestes barbarus* (Fabricius) (Zygoptera : Lestidae). *Odontologica* 13: 573 - 84

Waite, A. (2000) *The Kent Red Data Book*. Kent County Council.

Welstead, N. & Welstead, T. (1984) *The Dragonflies of the New Forest*. Hampshire and Isle of Wight Naturalists' Trust.

Further reading

Belden, P.A., Downer, V.J., Luck, J.C., Prendergast, H.D.V. & Sadler, D. (2004) *The Dragonflies of Sussex*. Essedon Press, Forest Row.

Brooks, S. (2002) *Dragonflies*. The Natural History Museum, London.

Cham, S. (2009) *Field guide to the Larvae and Exuviae of British Dragonflies. Volume 2: Damselflies (Zygoptera)*. British Dragonfly Society, Peterborough.

Silsby, J. (2001) *Dragonflies of the World*. The Natural History Museum, London.

Smallshire, D. & Swash, A. (2004) *Britain's Dragonflies*. Wild Guides Ltd., Old Basing.

Useful addresses

British Dragonfly Society, H. Curry, 23 Bowker Way, Whittlesey, Peterborough, PE7 1PT.
E-mail: hcdragon@btopenworld.com
Web: www.dragonflysoc.org.uk

Brook J.& G., 1 Barrack Cottages, Lower Street, Broomfield, Maidstone, Kent, ME17 1PU. BDS Vice County Recorders for Vice Counties 15 (East Kent) and 16 (West Kent). Kent Field Club Odonata Referees.

Kent Field Club, The Natural History Society of Kent, Hon. Sec., Ms. K. Friend, 2 West End Cottages, Doddington, Sittingbourne, Kent, ME9 0BZ.
E-mail: secretary@kentfieldclub.org.uk
Web: www.kentfieldclub.org.uk

Kent and Medway Biological Records Centre, Brogdale Farm Office, Brogdale Farm, Brogdale Road, Faversham, Kent, ME13 8XZ.
E-mail: info@kmbrc.org.uk
Web: www.kmbrc.org.uk

Kent Wildlife Trust, Tyland Barn, Sandling, Maidstone, Kent, ME14 3BD.
E-mail: info@kentwildlife.org.uk
Web: www.kentwildlifetrust.org.uk

Watkins & Doncaster (Entomological supplies), PO Box 5, Cranbrook, Kent, TN18 5EZ or visit the showroom at Conghurst Lane, Fourthrows, Hawkhurst, Kent.
E-mail: sales@watdon.com
Web: www.watdon.com

Glossary

Abdomen - the third section of the body of a dragonfly which consists of ten segments including anal appendages and genitalia.

Anal appendages - clasping appendages at the tip of the abdomen of adult dragonflies. The male uses these to hold on to the head or prothorax of the female prior to and during mating.

Antehumeral stripes - a pair of pale coloured stripes on the dorsal surface of the thorax of some dragonflies.

Basal spot/patch - a spot or patch of colour on the wing where attached to the thorax.

Costa vein running along the front or leading edge of the wing.

Costal band - a band of colour along the area of the costa.

Diapause - a period of suspended development at some stage of the life cycle (usually the larval stage in dragonflies) in response to conditions that become unfavourable for continuous development.

Dorsal stripe - a stripe of colour running along the central line of the upper surface of the abdomen of some dragonflies.

Dorsal spine - a projecting spine on the central line of the upper surface of certain abdominal segments of some anisopteran larvae.

Exuviae - the outer skin of the larva that is shed several times during larval growth. In the text this term is used with reference to the final shed skin that has been vacated during the emergence of the dragonfly. The Latin word has only a plural form, but an invented singular form, **exuvia,** is often used by odonatists for convenience and clarity.

Frons - the forehead or upper part of the face of the adult dragonfly.

Labium - hinged jaws attached to the underside of the head of the larva. These are used to catch and hold prey.

Lamella(e) - leaf-like gills attached in threes to the tip of the abdomen of zygopteran larvae.

Lateral bands/spots - pale coloured bands, stripes or spots on the sides of the thorax of some dragonflies.

Lateral spines - spine-like projections on the sides of abdominal segments of some larvae.

Maiden flight - the first flight of the adult dragonfly after emergence.

Node - that part of the leading edge of the wing where the costa is broken causing a slight bend or notch. Also that part of the lamellae of some zygopteran larvae where a slight step appears in the outline and where the setae may change in length or density.

Ovipositing - egg laying.

Ovipositor - apparatus for inserting eggs into plant tissue or other preferred material. Species which scatter their eggs have no ovipositor (see **Vulvar scale).**

Pre-flight - a newly emerged adult which has not made its maiden flight.

Primary genitalia - reproductive organs under the ninth abdominal segment of the male dragonfly from which sperm is transferred to the secondary genitalia. The primary genitalia of the female is under segment 8 and 9 sometimes with an ovipositor or vulvar scale.

Prothorax - a separate collar-like segment immediately in front of the thorax of adult dragonflies.

Pruinescence - a blue or grey powdery bloom on the abdomen and sometimes the thorax of some adult or over mature dragonflies.

Pseudo-pterostigma(ta) - white cell(s) near the tip of the leading edge of the wing of a female of the genus *Calopteryx.*

Pterostigma(ta) - distinctive coloured cell(s) near the tip of the leading edge of the wing, which acts as an inertial regulator of wing twisting.

Secondary (accessory) genitalia - reproductive organs underneath the second abdominal segment of the male dragonfly from which the female receives sperm.

Sentinel position - vertical position with wings closed taken by some adult male zygopterans whilst still clasping the prothorax of the female during ovipositing.

Tandem - a male and female dragonfly coupled together in flight or during ovipositing.

Teneral - a recently emerged adult dragonfly which still has a soft cuticle and has not attained full colour.

Thorax - the second section of the body of a dragonfly to which are attached three pairs of legs and two pairs of wings.

Vulvar scale - a backward projecting flap under segment 8 of some adult female dragonflies which scatter their eggs in the water.

Index of species names

Main entries are indicated in bold.

D

E

F

G

H

I

K

L